BERMUDA TRIANGLE SURVIVOR

Pilot Tells What He Experienced in the Heart of the Phenomenon

Bruce Gernon

Copyright © 2023 Bruce Gernon

All rights reserved. No part of this book may be reproduced or used in any manner without the prior written permission of the copyright owner, except for the use of brief quotations in a book review.

To request permissions, contact the publisher at
info@amazon-publications.com

Hardcover: 978-1-960657-02-2
Paperback: 978-1-960657-03-9

Edited by **Lynn Gernon**
Cover art by **Amazon Publications**
Layout by **Bruce Gernon**

Amazon Publications Group
1234 NW BOBCAT LANE,
ST. ROBERT, MO 65584-5678
Amazon-publications.com
+1 (646)-458-4222

Dedication

To my Granddaughter, Kali Burton, who has believed in my experience since birth and will pass on my theories to the future generations.

To my Grandson, Reed Burton, who I taught to fly at the age of five and one day we got to fly through a rainbow.

Table of Contents

INTRODUCTION 7

CHAPTER 1: FLYING A SMALL PLANE IN THE BAHAMAS 10

CHAPTER 2: THE SECOND BAHAMAS FLIGHT 14

CHAPTER 3: THE NEXT FIVE YEARS OF FLYING 18

CHAPTER 4: FLYING THROUGH THE VORTEX TUNNEL AND ELECTRONIC FOG 26

CHAPTER 5: A RISKY ESCAPE 37

CHAPTER 6: THE TUNNEL AND BEYOND 41

CHAPTER 7: ELECTRONIC FOG 52

CHAPTER 8: MODERN HISTORY OF THE BERMUDA TRIANGLE 58

CHAPTER 9: THE LOST PATROL 62

CHAPTER 10: THE GREATEST MODERN AVIATION MYSTERY 68

CHAPTER 11: THE ALIENS 73

REALITY CHAPTER 12: THE POWER OF THE ELECTRONIC FOG 97

ABOUT THE AUTHOR 101

INTRODUCTION

SURVIVED PILOT TELLS WHAT HE EXPERIENCED IN THE BERMUDA TRIANGLE

When I was a boy, I often had a peculiar recurring dream. I would dream I had the same abilities as Mighty Mouse and Superman. I would see myself flying all around the world—having the ability to do this with no visual means of propulsion and no wings. It felt great to leap into the air and fly. I could even hover whenever I felt like it. During the waking hours of the day, I would often try to fly with no luck. I thought it might be possible to levitate but that appeared to be impossible also. As the years passed by this dream continued. I started flying airplanes when I was seventeen and the dreams persisted. When I was in my late twenties, I started flying helicopters and when I learned to hover the dreams stopped and I have never had a dream that I could fly like Superman again. This made me feel much better, it was as if all my dreams and flying abilities had come true.

I have been encouraged to write this book because of the hundreds of students that have written to me and interviewed me over the past twenty years. There have been a vast number of older folks and scientists that have had an interest in my experience in the Bermuda Triangle over the past fifty years, but the number of young students has been three times the amount as them. Their interest has spread all around the world. The fascination with the

Triangle is based on one of the world's greatest mysteries, the threat of unknown forces, danger, and presumed disappearance into a portal or another dimension.

I am the only living person to fly through what I call a "timestorm" and experience a space-time warp, when I escaped through what I define as a "vortex tunnel." I am the only person to have seen what appears to look like a storm, from its birth stage, turn into a huge "timestorm" and fly through the center of it and escape through its "vortex tunnel." I have been researching this longer than anyone, ever since it happened in 1970 and I am the only living survivor that made it through the vortex tunnel in what is known as the Bermuda Triangle phenomenon. Due to my research, I have discovered something I call "electronic fog. I believe it is the key to the mystery and today's students will be the ones to prove it is real and eventually solve the mystery.

My story became known worldwide when it was featured in Charles Berlitz's bestselling book WITHOUT a TRACE in 1977. Starting in 1996 I have played a major role in over 50 TV documentaries and been on many dozens of radio podcasts. I co-authored THE FOG with award winning author Rob MacGregor in 2005. Our book has been published in seven languages. THE FOG has been revised under the title BERMUDA TRIANGLE LEGEND. Our latest book is BEYOND the BERMUDA TRIANGLE.

I was inspired to write this book because of two short videos that recently appeared on YouTube. One is a cartoon, and the other is a documentary, and they are both all about my experience. They have had over nine million views already, mostly by our younger generation. When the students read this book, they will gather much more information on what creates this mystery. This will complement the cartoons and documentaries giving them more knowledge to solve the mystery in our distant future. I have been researching this great mystery for fifty years—it may take another fifty years to finally solve it.

CHAPTER 1: FLYING A SMALL PLANE IN THE BAHAMAS

THE FIRST BAHAMAS FLIGHT

The seas were calm, the winds were light and variable, and the skies were perfectly clear. It would be an ideal day to fly to the Bahamas and back in the same day. I was only nineteen and had been flying for just over a year when I planned this trip from where I kept our airplane at Lantana Airport in Palm Beach County Florida. My two friends and I planned to fly to an airport in the small city of West End on Grand Bahamas Island. It was only 95 miles away and would only take about 45 minutes of flight time. We would be crossing over the Atlantic Ocean and the deep blue fast-moving Gulfstream. Little did we know that if our plans had worked out as we had hoped, we all might have died. You will find out at the end of this chapter why we almost faced a tragic death.

The year was 1966 and my dad had just purchased a new Piper airplane called a Cherokee Six. It was a beautiful red and white six passenger airplane powered by a single engine with a variable pitch propeller, and I loved it. My Dad had encouraged me to get a Private Pilot's license because he was a private pilot, and he knew I would be a natural at flying. There are two licenses you must attain to fly an airplane and take on passengers. First you must get the Solo license so you can fly alone or with your instructor and then the Private Pilot's license. I got my solo license after only six hours of instruction and my Private Pilot license after flying for a total of forty hours of flight time. Today it takes the average pilot about twice as many hours because airplanes and airspace are much more complicated now.

We brought along our SCUBA gear as we had all recently learned to dive with SCUBA (Self Contained Underwater Breathing Apparatus) tanks and were excited to dive in the crystal-clear waters of the Bahamas. We loaded up the airplane and departed early in the morning. We headed east and climbed up to 9,500 feet and leveled off while accelerating to our cruising speed of 165 miles per hour. We were flying VFR (Visual Flight Rules) that requires us to fly at an even altitude plus 500 feet when flying east and odd altitudes plus 500 feet when flying west. This is done for safety reasons to avoid head-on collisions with other airplanes.

After flying for 35 minutes, we spotted the island of Grand Bahamas when we were about ten miles east and descended into West End and identified the airport and made a perfect landing. What a relief! We were a little stressed out because this was our first flight into the Famous Bahamas Islands in a small airplane. We cleared customs at the airport Customs Office that was run by the British government. The local native people of the Bahamas took over the government from the British in 1970 in what they called a "Silent Revolution."

We explored the city but found there was not much going on in this small town. We asked if anyone could take us SCUBA diving, but it seems no one was available because anyone that had a boat was either out fishing or collecting Conch. Since we were unable to find a dive boat and it was getting late in the afternoon, I decided to head back to Palm Beach.

We departed West End and climbed up to 8,500 feet and started cruising to Palm Beach International Airport. We could not fly directly back to Lantana Airport because we had to clear customs at the nearest Customs Office first at Palm Beach International airport before landing at Lantana Airport, which was only five miles south of Palm Beach airport. I had only flown to the big international airport once before, so it made me a little nervous to go there. I was required to contact the Air Traffic Control tower on our radio when I was about ten miles out and requested to land. After landing they would switch me to another frequency called Ground Control and they would guide me to the Customs Office. The landing went fine and after the tower switched me to Ground Control, I asked them for "progressive taxi instructions" to Customs since I had no idea how to get there. We cleared Customs and then made the short hop over to my home airport at Lantana. It was a relief to depart Palm Beach International's controlled airspace and enter the more relaxing less controlled airspace of Lantana. It was a wonderful picture-perfect trip I will never forget.

Several months later a local pilot did something similar with his airplane and flew to Grand Bahamas and went SCUBA diving and then returned on the same day only he did not make it back to Palm Beach. He crashed in the ocean not far offshore from the airport. His airplane was located, and they did an analysis of him and found out he had the bends. It was later discovered you can get this decompression sickness if you go SCUBA diving and then flying on the same day, you must wait at least two days before flying. Back then SCUBA diving was still in its infancy and there were no rules or licenses required.

CHAPTER 2: THE SECOND BAHAMAS FLIGHT

My Dad and I worked together in the construction and development business. Dad had the aspiration to build a resort in the Bahamas and this was the main reason he had purchased our airplane. He told me that I was a much better pilot than him, so he made me the chief pilot. This meant that I would be flying the plane from the left seat which is designed for the person flying the plane. That meant I would be known as the "pilot-in-command" and be legally responsible for making all decisions and liable for the safety of everyone on board. Dad was a better navigator than me, so he was not only my co-pilot but the chief navigator.

Dad had arranged to meet with another developer that lived in Freeport on Grand Bahamas Island so I planned to fly with him when the weather was good, and the developer would be available. The year was 1967. I had been warned by some of my aviation mentors that when you fly to the Bahamas for the first times you may be fooled by an illusion that the clouds can create. If the clouds are scattered cumulus of medium size and you are flying above them, when you look down at the ocean you will see a shadow below the cloud that you might mistake for an island.

We made the arrangements to meet with the developer at the airport in Freeport in the morning and the weather was nice so it should be a good flight. The clouds were medium sized cumulus with no rain showers in the area. We departed Lantana Airport and climbed to 9,500 feet and began our cruise to Freeport which was 85 miles away. The flight time should have taken us about 45 minutes because there would be a 20 mile per hour head wind that would slow us down a little. After flying for about 35 minutes, we spotted the island of Grand Bahamas, so we prepared to start our descent for landing. After a closer examination we discovered it was not an island—it was a shadow!

Back then we had no weather avoidance equipment onboard the aircraft. The only weather detection equipment available then was radar and that was too expensive for a small plane. Today most small airplanes have lightning detection equipment and Doppler radar that shows radar images from a satellite. It was also difficult to determine your exact position, but it could be done by using our two electronic VOR's (omni directional equipment). We had to rely on what is called "dead reckoning" which is using your airspeed and compass plus the time in route to determine our approximate location. Today we have what is known as a "moving map" that always shows the exact location of the airplane.

We were fooled by a cloud. Since I was warned that this could happen, I was not too surprised, so we continued onward at our current cruise speed and altitude. A few minutes later we spotted the island of Grand Bahamas. At least we thought it was the island, but it was not. We were fooled again by the cloud shadows. A few minutes later we spotted the island again and again it was not the island. I was starting to become anxious. I felt like something was wrong. We should have spotted the island by now. I thought we had been flying close to arrival time, but I never checked to see what our exact flight time was, probably because I was becoming so nervous it was hard to think. I had the strange feeling that our airspeed was 165 miles per hour, but our ground speed was zero—we were flying but not making any headway. It felt as if we were standing still in space and time. Then it happened again—another shadow. I started to perspire and felt extremely uncomfortable. I thought we had been flying for over an hour and still had not found the island.

Finally, we spotted the island right below and in front of us. What a relief. I immediately felt much better. I began our descent at an increased rate as we were too high to land at 9,500 feet. We landed on the huge runway at Freeport which is 11,500 feet long. We landed at the beginning of the runway and were told to taxi to the other end of the runway, which was two miles away. While making this long trip on the taxiway I noted our flight time was 45 minutes. I could have sworn it took us more like 75 minutes. How could this be?

What happened to me is known as a time distortion. It is called a temporal time dilation. This transpires when time appears to slow down. It appears to be real, but it is all in your mind. It is usually caused because of extreme stress or anxiety. The person it happens to can usually remember every second of the event and they are extremely aware of how long it seems to be taking. I can remember every minute of this event flying to Freeport and how it seemed to be taking way longer than it should have been. I met a retired marine combat soldier once while doing a TV documentary and he told me what it was like when he was in a combat vehicle when they ran over a buried bomb and it exploded, and he lost his leg from his knee down. He said that while he was waiting for someone to help him, time slowed down so much that one second felt more like one minute. He thought it took the medics an hour to rescue him, but it was only a minute.

There is another form of a time distortion that is just the opposite of time dilation, it is called time compression. Time seems to speed up when you are in this situation. Some people think this is what I experienced when I made my famous flight through the Bermuda Triangle and had a space-time warp, but that is not what happened. When someone experiences a time compression there appears to be missing time and they cannot remember where the time went. They cannot remember what happened while the time was missed, everything during that time was a blur. I can remember every second of that flight. Nothing was blurry except when I was in something, I later discovered that I call electronic fog. I was in it for only three minutes, and I believe this electronic fog is what causes a major part of the Bermuda Triangle mystery.

CHAPTER 3: THE NEXT FIVE YEARS OF FLYING

Over the next five years Dad and I would fly to the Bahamas at least once a month looking for land that would be a good location for developing a resort. We traveled to most all the major islands in the Bahamas and even went as far as the Turk's and Caicos Islands just beyond the Bahamas. It was exciting and always an adventure going to a new location. We would often fly to Nassau, the capital of the Bahamas, to meet with our attorney and do development research for the company we had formed there.

A pilot is always required to get a weather report from a meteorologist before flying anywhere beyond their local airport. Back then the weather reports were not nearly as accurate as today. Today you can get a certified weather report from a computer. Predicting the weather has never been an exact science and it is one of our youngest sciences. Forecasting the weather has improved a great deal recently, especially with hurricanes. We would always only fly when the weather was good but there were several times when we would encounter a large squall while we were in route and had to turn back to where we had departed from. Over the next five years I would accumulate around 100 hours of flight time per year and fly to the Bahamas almost one hundred times.

FLYING FOR FUN

One of the memorable trips I took to the Bahamas took place around 1968 when I flew five of my boyfriends to Nassau in my Cherokee just for fun. The main reason we were going was to witness the huge celebration they have on the island every New Year's Day called the Junkanoo. One of my friends I had known since junior high school, named John Woolbright, had graduated from college with degrees in mathematics and had become a research scientist for A.U.T.E.C. on Andros Island. This is a research facility for Navy submarines. We landed at the Andros Fresh Creek airport and picked up John where he lived on the Navy Base and then flew on to Nassau International airport.

We spent a day and the night there and while we were there, we got into a disagreement. It was a friendly type of intelligent argument over the question of whether we can produce zero gravity on the surface of Earth. John and I and one other friend believed it could not be done. We believed the closest thing to it is SCUBA diving underwater or forming a parabolic curve in an airplane. My other three friends thought the government had developed a zero-gravity machine. We ended the night undetermined as to who was correct about the question of zero-gravity.

The next morning, we went to the airport and planned to fly direct to the Andros airport to drop off John and then proceed back to Palm Beach. It was a 30-mile trip from Nassau to Andros and it would be flying over what is called the "Tongue of the Ocean." This is an area almost 30 miles wide and 100 miles long and 6,000 feet deep that runs between Nassau and Andros Island. This also is the exclusive area for the Navy to test their submarines and torpedoes.

When we were about halfway to Andros, near the center of the Tongue of the Ocean, my friend John yelled out loudly from the back seat, "Gernon give me zero gravity!" I started thinking, well we are cruising at 2,000 feet and just below the ceiling of some medium sized cumulus clouds so I could dive down and pick-up speed and then shoot up into one of the clouds and show them zero-gravity while inside the cloud. I said, "Okay tighten your seat belts." I dove down and picked up a speed of close to 200 M.P.H. and then shot upward and timed it exactly right so we would be inside the cloud when I would enter the top of the parabolic curve and experience zero-gravity. It worked out perfectly and we all experienced zero-gravity for about seven seconds while the visibility went down to near zero. Everyone was screaming except me. I was laughing my head off.

When we popped out of the cloud all we could see was the deep blue waters of the Tongue of the Ocean and everyone was holding their breath and gasping as we gently pulled out of the steep dive skimming just above the surface of the ocean. It was the first time any of them had ever experienced zero-gravity and to do it in such an inconceivable and remote location made it even more exciting. The rest of the flight was uneventful as we dropped off John at Andros Island and then flew on to Palm Beach International and cleared Customs.

DEVELOPING AN ENTIRE ISLAND

In 1969 I worked for a real estate and development company based in Delray Beach, Florida. They had a development project in the Bahamas and were selling lots in their subdivision located on Great Exuma Island. They had a partner that lived on the island named Fritz Ludington. He wanted to develop another subdivision on a British government owned group of islands just beyond the Bahamas called the Turk's and Caicos Islands. One of the islands near Caicos is named Providenciales, also known as Provo, looked pristine and had never been developed. Fritz made an appointment with the governor of the islands on the capital island of Grand Turk, and he wanted us to come with him on the day they would meet.

We took my airplane along with two real estate associates and flew from Palm Beach the 300 miles directly to Great Exuma. We met with Fritz the day before his meeting with the Governor. We spent the night at Fritz's hotel named the Two Turtles Inn. The next morning, we met at the airport and Fritz took his airplane along with three of his associates and I took mine along with my two associates and we flew the 350-mile trip in formation directly to Grand Turk Airport.

When we reached the chain of islands Fritz pointed out Provo to me. We circled around over the island, and I could see how magnificent it was. It is about 15 miles long and five miles wide. It has an expansive shallow bank on the west side and the deep Atlantic Ocean on the east side. There were many miles of sandy beaches. It has a barrier reef that runs parallel to the east shoreline that starts at around 100 feet deep and drops off to 2,000 feet and had never been explored by SCUBA divers before. We flew around the interior of the island and noticed there were two freshwater lakes that were loaded with hundreds of pink flamingos. It was an astounding island ripe for development. From there we went onward to the international airport in Grand Turk Island.

We both landed at the airport and then went to our hotel and checked in and had dinner there. While we were there, I was honored to meet the famous airplane designer Burt Rutan and his wife and his brother Dick Rutan. They had flown his twin-engine Piper Apache airplane all the way from California and were in route to South America. I got to talk to Burt for a long time before we all retired for the night after having a long day of flying.

Early the next morning we all got up and had breakfast and then went downtown to meet the governor at the capitol building. Fritz went in alone and we all stayed just outside near the building in the center of town. We waited for Fritz and after about an hour he came out of the building to greet us. We did not know what to expect and hoped the governor was willing to work on a deal with him. What happened next, I will never forget. Fritz started jumping up and down and yelling and singing and dancing a jig right in front of us in the middle of the street in the center of town right in front of us and several local towns' people. We were all in shock and wondered what had happened. He finally came out and said over and over, "You are not going to believe this—he gave me the whole island! He gave me the whole island" . . .

Now we were really in shock. Could it be possible? Did the governor really give him the whole island to develop at no charge? We would find out later that he did not quite give him the whole island, but he did give him his choice of any location and 4,000 acres to go with it. That is still something incredible for a government to do for someone. Fritz was in heaven as we drove back to the airport to head back to Exuma.

When we arrived back in Exuma we went to Fritz's hotel, and he had a party there with free food and drinks for us and everybody on the Island. Everyone was drinking and partying so I went to bed early because I knew I would be flying early the next day. I can remember hearing Fritz's loud voice, above all the racket from the party, while trying to go to sleep, laughing and still repeating that they gave him the whole island, until the wee hours of the night.

Fritz went on to develop the island, creating a subdivision and a resort hotel and marina called the Three Turtles that became a popular destination resort. Today Provo is an international destination with an international airport and many resorts. It has the largest population of all the Turk's and Caicos islands. I returned there after being away for twenty years and I was astonished to see how it had expanded. I picked up a locally published magazine and it had an article in it about how it all started in Provo in 1969. I was amazed to read the story because it explained to me why Fritz was so excited and lucky when the governor gave him the land.

The magazine had a story about Fritz and a preacher. The preacher was born and raised in Provo in the only tiny village on the island. He was beloved by all the natives that lived in the Turk's and Caicos Islands. In 1968 he had a vision that seemed immensely powerful to him. It was a vision of a man that would come to the Turk's and Caicos and meet with the governor and ask him if he could develop Provo. The man would go on to develop Provo into the most important and populated island in their chain of islands. It would be wonderful for all the natives on these islands because they were poor and mostly unemployed and desperate for work.

The preacher met with the governor and told him about the vision. He told him a man would come to Grand Turk in his own airplane and meet with the governor. He would be a white man that is tall and tanned with short grayish brown hair and a short full beard—this is what Fritz looks like. He would speak with a slight foreign accent—Fritz was from Germany. The preacher told the governor, "When you meet with this man, give him whatever he asks for because it will be wonderful for the future of Provo and all the Turk's and Caicos Islands people."

Within a year later Fritz met with the governor and the rest is history. No wonder Fritz and all of us were shocked on the day he met with the governor, and he gave him all the land he wanted. That was an incredible vision the preacher had, and it must have been incredible for Fritz too. Unfortunately, he died young and never got to see the amazing expansion of the island. I feel fortunate to have seen this miracle even if I only played a small part in it.

BRAND NEW AIRPLANE

Early in 1970 we purchased a brand-new Beechcraft Bonanza A36. Beechcraft is famous for creating and producing this aircraft back in 1948. It was ahead of its time because it was fast and had solid flying characteristics. It was well-known for having a V shaped tail instead of the standard triple tails. The A36 was the first of its type to come out in 1970. It was a single engine propeller airplane like the V-tail but it had three tails and a more powerful engine and was stretched one foot longer so it could accommodate six people. It was a pleasure to fly this airworthy airplane. Because this airplane is so well made it is still in production today after 51 years. A new one today has all the modern advanced avionics, but the airframe is nearly the same as it was in 1970.

(Pre-Flighting our Bonanza before flying back home)

We finally located an island that was situated just offshore of the island of Andros that would be perfect for developing our resort. Andros was located not far from Nassau and is the largest island in the Bahamas. We started flying there around once a week because we were preparing to build an airport on our private island. I had accumulated just over six hundred hours of flight time and flown back and forth from Palm Beach International airport to the Andros airport at least a dozen times prior to the flight that changed my life on December 4th, 1970.

CHAPTER 4: FLYING THROUGH THE VORTEX TUNNEL AND ELECTRONIC FOG

THUNDERSTORMS

My dad and I along with our business partner, Chuck Lafayette, spent the day working on Andros on Thursday and were going to head back to Palm Beach on Friday. That night I noticed the weather was so clear I could see all the stars and planets and even the Milky Way. We planned to depart Andros Island first thing in the morning, but just before the Sun came up while I was still in bed, I could hear the crackling sound of lightning along with seeing the room light up from its flash and then the loud sound of thunder. I knew we would not be departing early. In 1970 telephones did not exist on the island so I could not get a weather report so we packed our bags and headed to the airport hoping it was only local thunderstorms and they would soon dissipate. We never made it to the airport because the thunderstorms were too intense and appeared to cover most of the island of Andros so we waited at our hotel hoping they would eventually dissipate.

Finally, at around 2:00 PM the thunderstorms let up and there was no more lightning, only steady light rain and an overcast sky, so we drove to the airport. I gave the airplane a pre-flight and we taxied to the runway for takeoff. Our plan was the same as always. We would fly directly over Bimini Island and then turn slightly northward and fly directly to Palm Beach International. The flight distance would be 210 miles and take approximately 75 minutes. We lifted off the runway at exactly 3:00 PM and I set the airplane's timer and synchronized our wrist watches. We climbed up to 1,000 feet and leveled off and established a cruise speed of 180 MPH. It looked like the entire island was overcast at 1,500 feet and the light rain was continuing. We turned to a compass heading that would take us directly to Bimini as we crossed over the interior of Andros.

The light rain continued while cruising for almost ten minutes, but the air was stable, so we continued while maintaining an altitude of 1,000 feet. We would not be able to get a weather report from Miami until we reached a higher altitude. While we were approaching the northwestern shoreline of Andros, I could see the clear aqua waters of the Great Bahamas Bank and noticed that the overcast storm clouds ended at the shoreline, and it looked like there was clear blue sky ahead.

When we got close to the shoreline, I noticed there was a strange looking cloud about a mile offshore that was directly in front of our flight path. It was hovering over the ocean at about 500 feet and was about a mile long and 1,000 feet thick and 1,000 feet wide. It appeared to be a lenticular cloud. This type of cloud is named this because it has the shape of a lens from a telescope. They are oval shaped and have smooth silky edges. They normally only form at high altitudes near mountains. I have seen them in south Florida only a few times and they were all at high altitudes above 10,000 feet and moving rapidly with the wind. This cloud was not moving at all and was much lower than a normal lenticular cloud. It had an odd cigar shape to it and seemed to be out of place.

(A rare lenticular-shaped cloud appears directly in front our flight path hovering just offshore over the Great Bahamas Bank.)

The cloud appeared to be motionless and harmless, so I continued my flight path and pulled the wheel back and started climbing right over it to my designated altitude of 10,500 feet. When I reached an altitude of 3000 feet, I was able to contact Miami Flight Service on the radio. I filed my required flight plan with them and asked for a weather briefing. They told me the weather was clear between Andros and Miami and there were a few scattered thunderstorms around south Florida and the winds were light and variable. Everything was looking good until I went inside a cloud that seemed to come from beneath our airplane.

Inside the cloud It appeared to be like a normal cumulus cloud with visibility only about fifty feet and pure white in color—but where did it come from? I was inside it for about thirty seconds and then I popped out of it, I could see the cloud was spread out all around us and underneath us. When I got about 200 feet above it, I did not stay there long because the cloud caught back up to us and I had no choice but to go back inside it.

The only thing I could figure out about what was happening was when I flew over that strange oval shaped cloud it just happened to be in the birth stage of a cumulonimbus thunderstorm and it was building up right underneath us. We were climbing up at 1,000 feet per minute and it must have been climbing up at a faster rate than that because whenever I went inside it appeared to boost me out of it but then it would catch up to us and we would go back inside. I must have gone inside and out of the cloud at least five times. When we reached an altitude of 10,500 feet and my dad suggested to me that we should turn around and go back to Andros. When you get caught inside a storm it is usually a wise decision to make a 180 degree turn and get out of the storm. It was a difficult decision, but I decided to continue because I didn't like the idea of flying back into this building storm. After only one more minute, we shot out of the cloud at 11,500 feet and saw clear skies ahead. I leveled off and went to cruising speed of 195 MPH and when I looked behind us, I could not believe what I was seeing.

(Relived to finally escape this storm build up I was amazed when I looked behind me and on both sides at how much the storm had expanded)

The cloud was much wider and higher now. It looked like it was nearly 20,000 feet high, and it was in the form of a half circle, like a horseshoe. It was spread out on either side of me as far as I could see, which was at least twenty miles. It looked like it was not just a thunderstorm, it was like a squall line in the shape of a semi-circle for as far as I could see. There was something odd about the shape of it. Squalls usually have uneven scalloped edges and curves as they spread out. This cloud appeared to have an edge that formed a perfect semi-circle with no uneven edges kind of like looking at the inside at the eyewall of a hurricane. Another strange thing I didn't realize until much later, was the fact that while I was climbing through the cloud our airplane was traveling at 105 MPH and the cloud was spreading out faster than that. How could that be?

After the cloud appeared to stop spreading out in front of us and we left the cloud behind and continued under clear skies toward Bimini. I engaged the autopilot, established a cruising speed of 195 mph, sat back, and started to relax. Everything was back to normal, or so I thought. I tried to forget the strange experience we had just had. I maintained our altitude at 11,500 feet. For some reason I felt safer at a higher level than planned. The higher the altitude the safer you are in case of an emergency because you can glide further towards the nearest island. I did not realize that the challenge was just beginning, and we had not escaped at all.

After another few minutes of flying in the clear skies and moving closer and closer to Bimini, we noticed another squall directly in front of us. As we approached the cloud, moving at just over three miles a minute, an eerie sight unfolded. To my dismay, the cloud looked very much like the one we had left behind, it had a similar smooth curving, semi-circular shape, except the arms extended in the opposite direction, as if attempting to embrace us. This cloud was even taller than the one we had just escaped from with its top reaching at least 40,000 feet.

Then I noticed something else that stunned me. Normal cumulus clouds have a base or ceiling 2,000 to 3,000 feet above the surface. If the cloud is producing heavy rain, the base is usually about 1,000 feet and sometimes as low as 400 or 500 feet. But, as we flew within a few miles of the cloud, I saw that this cloud appeared to rise directly from the ocean. This was another highly unusual development. Cumulus clouds do not lie on the surface of the ocean, they always have a base or ceiling above the surface. Fog banks are known to lay on the ocean, but they stay at lower altitudes and never reach 40,000 feet.

I realized that we could not go under the cloud or above it and attempting to fly around it would take us considerably off our flight path. Besides, the arms of the cloud were already stretching out on either side of us, so we could not make an easy escape. However, the cloud did not look too threatening, so after conferring with my dad, I decided to fly into it while maintaining our direct course to Bimini.

I had flown under clouds in heavy rain, and I had penetrated into them when they were not storm clouds and had low visibility. But pilots are supposed to steer twenty miles clear of strong thunderstorms, and the 10,000-foot level was supposed to be the most dangerous altitude to fly through a storm. I had been told that there could be updrafts and downdrafts of more than 100 MPH in the heart of the thunderstorm cell.

We were about 45 miles east of Bimini when we entered the misty edges of this enormous cloud formation. Once inside, I realized I might have made a mistake. Although the cloud was white and silky on the outside, the interior was dark, as if night suddenly had fallen. But it did not stay dark for long. Bright white flashes lit up the interior of the cloud. They seemed to go on and off in a never-ending, random pattern, and the deeper we penetrated, the more intense the flashes became. It seemed strange that there were no lightning bolts, and the air was stable.

Even though there were no bolts of lightning, and no rain, the air was not turbulent, and visibility appeared to be at least two miles because every time it flashed, I could see the ocean below and beyond. I had no doubt that we had entered an electrical storm of an extremely high intensity. We were in trouble, and I was getting more concerned by the second. I was starting to get a bit nervous when my father asked if I was going to continue. I shook my head—no—and immediately turned left to a southerly heading and darted back out of the storm.

(Heading south we flew along the edge of the storm and noticed it was curving back toward the west where the storm we had just escaped from was located)

Dad and I checked our watches and noted that we were deviating from our planned course at 3:27 PM. The mechanical-powered clock on the panel, which included a timer that I had engaged upon takeoff, indicated that we had been airborne for 27 minutes. When we changed course, my father started the timer on his watch, and by using the plane's electronic navigational equipment, he calculated that we were 40 miles southeast of Bimini. Meanwhile, I contacted Miami Air Traffic Control (ATC) and told them that we had deviated from our course to avoid a thunderstorm and were attempting to fly around it.

Surprisingly, the air traffic controller thought it was no big deal to fly around this storm and calmly told us that should be no problem and to contact him after we got around the storm. Thinking about his attitude today I believe he was not picking up any major storms over the Great Bahamas Bank on his radar or he would have warned us there could be a problem trying to fly around the storm. Not only that, but he had also told us 15 minutes earlier there were no storms between Andros and South Florida. Is it possible that radar could not pick up this storm?

We were still concerned about our situation, but we thought we might be able to avoid the semi-circular-shaped cloud to the south by flying around it. However, after traveling six or seven miles we could see that the cloud continued on our left to the east. A minute later, we realized we were in serious trouble. Astonishingly, we could see that the cloud we had encountered near Andros was now connected with the second storm cloud. As far as I could tell, the enormous cloud encircled us. I estimated that the diameter of the clearing within it was about 30 miles. We were trapped inside a huge donut hole, a billowing prison with no way out. We could not fly over it. We could not fly under it.

CHAPTER 5: A RISKY ESCAPE

Now I was really getting worried, but I knew I needed to remain calm. When I tried to understand how we had gotten into this predicament a stray thought came to mind: I remembered my mother's story about when I was born during the largest thunderstorm she ever witnessed. Now here I was inside what might be the last thunderstorm I would ever encounter.

It seemed that the storm was created initially from a saucer-shaped cloud just offshore of Andros Island. It had rapidly spread outward, forming the horseshoe shape, then it connected on the opposite side forming a circle about 40 miles away from where it was born. I recalled what it was like inside the thunderstorm, and I did not want to fly back into the storm with all those lightning flashes.

This storm had spread out at a phenomenal rate. It had built upward to over 40,000 feet in less than 15 minutes and it had stretched horizontally which I estimate at over 200 miles per hour. The tops of the storm had the rare silver lining on the edges that can be seen when the tops are high, and the sky is clear, and the sun is low. I believe it would not be a 200-mph wind though, it would be more like an ignition. It would be like pouring a circle of gasoline and igniting it on one end then it would rapidly spread to the opposite end forming a circle of fire.

We had flown about ten miles from the point where we turned off course to the south when I noticed an opening in the massive cloud. Apparently, the squall shaped storm had formed near Andros then spread out forming a circle and when meeting on its opposite end it came together on the bottom first forming what looked like a valley between the two ends. At the top of the cloud, on either side, arms started extending outward above the valley forming what is known as a thunderstorm anvil-head. It looked as if the arms would soon connect, creating a bridge and forming a tunnel.

(Trapped inside the torus-shaped squall with no way to fly under or over, that formed into a tunnel. I felt it would be our only escape, so we decided to fly through it.)

The anvil shape is commonly seen in cumulonimbus thunderstorms when the winds are light as they reach maturity. The top typically spreads outward and horizontally for several miles at around 35,000 feet. Normally, I would stay clear of any anvil head-shaped clouds because there could be hail underneath them, but our situation required drastic action.

Faced with the predicament we were in, I felt that I had no choice but to turn the aircraft to the right and assume a heading of 290 degrees aiming directly for the exit through the cloud by way of the only visible opening. I consulted with my father as to whether we should take a chance and fly through the opening, and he concurred with me that it was the only way out. As we flew toward the aperture, we saw the two anvil heads connect, forming a tunnel that was about a mile wide and appeared to be between ten and 15 miles long. Its center was at an altitude of approximately 10,000 feet. On the far side of the passage, we could see a clear blue sky. That gave us hope.

As we neared the tunnel, we realized that its diameter was shrinking. So, I took the engine up to maximum power. By the time we were two miles away, the opening was only about 1,000 feet wide. When we were one mile away, the opening was only about 700 feet wide. I recalled what my first flight instructor, Charles Galanza, told us in class one night, he said to never fly through what he called "sucker holes." They usually form between thunderstorms about two miles high and can be miles long. He had a friend that flew through one and disappeared. I thought he meant that his friend crashed. Now I believe he meant he disappeared inside the tunnel.

We were still one-half mile away when the opening shrank to 500 feet. The center of the tunnel appeared to be at 10,000 so I started descending from 11,500 feet and watched the air speed indicator increase to the never exceed red line at 230 MPH. We were now committed to enter the tunnel because making a turn at this speed would cause structural damage to the airplane. As we approached the opening, the aperture was about 300 feet wide and still shrinking. I leveled the airplane at 10,000 feet and prepared to fly through the tunnel. We entered the tunnel at exactly 3:30 PM.

CHAPTER 6: THE TUNNEL AND BEYOND

As soon as we flew into the tunnel, I was startled by the strange spiraling lines hovering a few feet along the interior of the tunnel walls. The walls of the tunnel were smooth and had that silver lining color, glowing while the spiraling lines looked like puffs of gray clouds varying in size from about ten feet to thirty feet or about the size and shape of a school bus. They broke apart from each other forming a line that swirled all the way to the end of the tunnel while they slowly rotated counterclockwise.

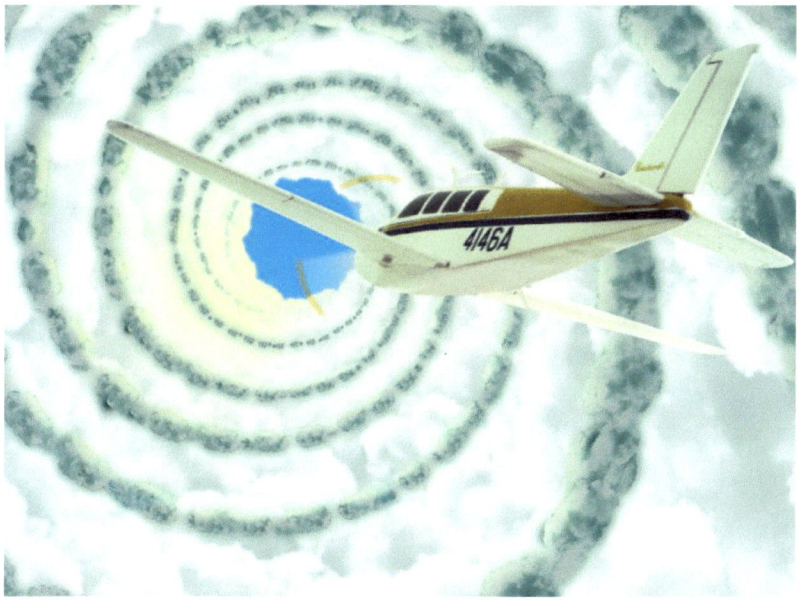

(Inside the vortex tunnel I` witnessed something phenomenal taking place. Was I seeing the creation of the electronic fog?)

Before we entered the tunnel, we could see clearly to the other end and there were no swirling lines. Why did they form instantly upon the emergence of the aircraft? No scientist has been able to answer this question so far. Recently I have had some amateur scientists contact me with their theories. One even sent me a photo taken in his lab that showed something similar looking to the tunnel I flew through. He said that I experienced a frictionless motion that he called "aqua planning." Another said I was experiencing a gravity wave, so, I must have been aqua planning on a gravity wave. It could be possible that when the two ends of the Timestorm anvil heads came together there were positive electric charges on one side and negative charges on the other side and a dipole field was created within the tunnel, creating an electromagnetic energy force field in an east-west direction. I believe what I saw was the beginning of a space-time warp and I was seeing the fabric of time itself while I was flying through the tunnel. I also believe this is where I saw the formation and creation of my theory of electronic fog.

Seconds before I entered the tunnel, it appeared to be at least ten miles long. Now it appeared to be only about a mile long and we could still see blue sky on the other side. Instead of taking the three minutes it would normally take to travel the distance of the tunnel it only took 20 seconds to pass through the tunnel. I concentrated on trying to remain right in the center of the opening. I was afraid that if the wings brushed the edges of the cloud, I might lose sight of the hole on the other end and the path to the clear sky. The silky white walls of the tunnel glowed with the light from the afternoon sun, and the tunnel was still shrinking. The diameter of the tunnel had reduced to only 30 feet when the tips of the wings scraped the edges of the cloud as we reached the far side and escaped the collapsing tunnel.

(Exiting the tunnel was the point of attachment of the electronic fog. I felt a strange sensation like I was floating and skidding forward through space much faster than normal)

That is when I suddenly felt an unusual weightlessness and knew my seatbelt was the only thing keeping me from rising out of my seat. It was the only time I have ever experienced that sensation in my 53 years and 3,000 hours of flying. It felt like zero gravity, but it also felt like I was hydroplaning at the same time. I felt like I was going forward much faster than normal, like I was aqua planning on a gravity wave. While this was happening, I noticed vapor trails at the ends of the wing tips that created parallel contrails behind us leading to the tunnel. After about ten seconds, the weightless sensation vanished, and I felt disoriented because visibility had instantly changed. The clear blue sky was gone, and everything looked foggy. It was at this point where what I call electronic fog attached itself to our aircraft. The electronic fog must have attached itself to us while we were inside the tunnel because we could see the clear blue sky outside of the tunnel, but it instantly disappeared when we exited the tunnel.

I looked back and gasped as I watched the tunnel walls collapse and form a slowly clockwise rotating slit. I was relieved that we made it through, but something seemed to be wrong with our instruments, so I asked my dad to check our position. He was always good at using the instruments to give me our exact location on the chart within only a few seconds. This time he fiddled with the instruments for longer than usual. Then he said something was wrong. That was when I realized that all the electronic and magnetic navigational instruments were malfunctioning. Even the magnetic compass was slowly rotating counterclockwise, as if the plane were making a turn.

Now, instead of the clear blue sky, everything appeared a dull grayish white with a tint of yellow. Visibility appeared to be at least two miles, but there was absolutely nothing to see—no sky, no horizon, only a grayish yellow haze. I could make out the ocean below us, but it did not look right, it was blurry and fuzzy as if electricity were somehow mixed in the haze. Although haze in the lower atmosphere is common, this haze was darker than normal and like nothing I had ever seen before. Not feeling comfortable flying through these conditions, I slowed down to the safe maneuvering speed of 180 mph. We seemed to be in some sort of fog, but unlike the usual fog where visibility is never much more than a few hundred feet, we could see much farther. Even more disturbing was the fact that the instruments continued malfunctioning.

(We flew with the electronic fog attached to us for three minutes)

I was able to contact Miami air-traffic control and reported that I should be ninety miles east of Miami but was not sure of our position and would like radar identification. The plane was equipped with a transponder, a new invention for small planes in 1970 that helped radar controllers identify the location of airplanes. I told them that we were about 45 miles southeast of Bimini heading west and flying at 10,000 feet. But the controller came back and said there were no planes on radar between Miami, Bimini, and Andros. Even our new transponder appeared not to be working. That was when Dad snatched the microphone from me and yelled at the controller in a scolding voice. "What the hell do you mean you can't find us on radar?" That was the first and only time we had ever cussed over the radio.

It was a terrifying moment. We literally did not exist, at least not on radar. Had we crashed and died, and did not know it? I did not think so. But it was probably fortunate that I did not know anything about the Bermuda Triangle—none of the bizarre stories—because I was inside of one!

The controller sounded bewildered and apologized, but said the radar showed no blips in the area we were flying. I wondered how that could be. In the past, they had always been able to identify us, especially when we were approaching the Air Defense Identification Zone (ADIZ). This is an imaginary line that runs parallel with the Florida coastline and is about five miles offshore. All airplanes must be identified with Air Traffic Control before penetrating it and entering Florida airspace.

Dad was getting more and more agitated and began screaming some more at the controller. He was starting to panic so I grabbed the microphone back and was able to calm him down by telling him everything would be alright.

But I was wrong. It was about to get much more baffling and disturbing. Chuck spotted an island below and thought it might be Bimini, but on closer examination I decided it was just a shadow from a cloud. According to our flight time Bimini should still be 35 miles ahead of us. The shadow was hard to make out through the haze and it seemed to pass by us at a much faster rate than normal. Chuck continued to talk to me, but I could not understand a word he was saying. Something strange was happening to him and he was distracting me, so I had to totally ignore him and concentrate on flying the airplane. Later, he would tell his children that the hands on his watch were spinning as we flew through the electronic fog, but I think it only appeared that way to him because of his state of mind.

DISLOCATION

I continued cruising at 180 mph, because I did not know what would happen next. I remembered that when we entered the tunnel our heading was 290 degrees, but now the compass was continuing to spin. Very soon we could be going in any direction, including right back into the dangerous cloud. Desperation and anxiety crawled across the pit of my stomach. I focused on an imaginary compass I had created in my mind that always pointed to true north. I had worked on this form of navigation for many months just before this flight and had positive results. This would be the first time I needed to use it. I set the imaginary compass to a 290-degree heading and maintained that course.

At this point, we had been traveling for nearly 32 minutes. According to our flight time, we should have been approaching the chain of Bimini Islands, which extend 55 miles to the south of Bimini, the main island, to Orange Cay, the southernmost island in the chain. I estimated that we were about 85 miles southeast of

Miami, and still looking for the Bimini chain of islands. If my internal compass was working, I figured we would be crossing the islands in six or seven minutes. If it was not working, well, we would not be heading anywhere near Miami. That was for certain.

We continued, still shrouded by the odd haze. I was puzzled by the conditions, but the air remained stable, and I felt in control of the airplane. We were still on the same frequency but had not heard any transmissions over a microphone. That seemed odd, but it was probably because there were many airline pilots flying in our area, and they were being silent because they were concerned about our outcome. Then, suddenly, the controller yelled out that he had spotted an airplane directly over Miami Beach, flying due west. He seemed to be overly excited about it.

I looked at my watch and saw that we had been flying for just over 33 minutes since liftoff. We could not possibly be over Miami Beach already, we were approximately 80 miles southeast of Miami, and still looking for the Bimini chain of islands. He seemed upset that it was not us and asked, "Are you sure you are not over Miami Beach?" I replied, "Affirmative, we should be about 80 miles east of Miami."

Suddenly, the fog started to break apart, but it did not just dissipate. Long lines in the fog ran parallel along our direction of flight and all around the airplane. The lines slowly cracked open leaving ribbons of fog spreading apart about one mile from the plane and were running two to three miles in front of us, and about the same distance behind us. The slits gradually grew wider leaving ribbons of fog that slowly got skinnier until they disappeared leaving only clear blue sky. It only took about ten seconds for this event to happen. I call it "electronic dissipation."

(Another incredible event took place when we reached Miami Beach and the electronic fog dissipated. Again, if you watch my videos the event will be more definitive.)

When I looked behind me all I could see was a bright blue sky with the visibility about ten miles and no storms. How could this be? I was expecting to see a huge fog bank but there was none. I would later figure out that the dissipation created an illusion in my mind. The ribbons that appeared to be miles long were only yards long. Maybe they were only one hundred yards or so. I wonder if I was seeing an illusion when I flew through the tunnel vortex? If it was an illusion, what did it really look like?

As my eyes adjusted to the brightness, we were astonished to see Miami Beach directly below us. We were relieved to see familiar land again and to have escaped the fog. But how did we get here so fast? Just three minutes and 20 seconds ago we were 100 miles from Miami and now we were directly over Miami! We had passed over Bimini without seeing it and we had passed over the ADIZ without getting the required permission. Something did not seem right. To accomplish this, we would have to be traveling at nearly 2,000 mph.

I contacted the controller and told him he was right; we were directly over Miami Beach and still at 10,000 feet. I told him we would sign off and continue onward to Palm Beach International. He was reluctant to sign us off, probably because he realized something supernatural had happened so I told him we would be fine and said, "thank you—good day." Then another supernatural thing happened to me. A powerful voice said to me "Never forget what you have just witnessed because someday it will be important for the whole world to know about." Was that my Guardian?

Angel telling me this?

Dad noticed the electronic instruments were working again so I turned to the north and started our descent into Palm Beach. Chuck appeared to be back to normal. It seemed that the electronic fog influenced his mind, and he recovered shortly after it detached from us. While in route we encountered a large thunderstorm over Ft. Lauderdale and had to deviate 30 miles to the west over the Everglades to get around the storm. After flying around the storm, we headed to the northeast and landed at the airport and cleared Customs.

When we touched down on the runway, I noticed the airplane's timer indicated that the flight time was only 47 minutes. This could not be right. I checked my wristwatch, and it indicated the time was 3:47 pm. Dad's wristwatch indicated the same time. This did not make any sense because I had made this flight many times and it always took about 75 minutes on average. I was expecting this flight to take even longer because it was not a direct route. We had to deviate about 40 miles out of the way. Our normal route was 210 miles long, this flight was closer to 250 miles and should have taken more like 85 minutes.

I had no answers as to what we had just experienced but I realized that something incredible had just happened. Something so significant that someday it would be important for the whole world to know about. I did not realize that this would be the beginning of a lifelong quest to find the answers.

CHAPTER 7: ELECTRONIC FOG

After that flight I did not talk to anyone about it, not even my dad, because it was unexplainable. Because I believed something important happened on the flight, I decided to commit it to my memory. Every day I would review the flight in my mind at least once or twice a day. After the months passed by it started to become a permanent visual part of my memory. Then almost exactly one year later I learned from a scientist named Ivan Sanderson, there was an area in the Bahamas and part of the Caribbean where boats and airplanes have mysteriously disappeared, and he believed it had something to do with time.

When I heard this, it was as if a light bulb had gone off inside my mind. The fabric of time would be the answer to what I had experienced. Sanderson called this area "the limbo of the lost" and it later would become known as the Bermuda Triangle. After learning this, I spent a great deal of time in my college library, reading everything available about this great mystery and researching everything I could about the subject of time.

As the years passed by there was one part of the flight that I could not understand. After I came out of the tunnel what happened to the clear blue sky? I thought I had entered a fog bank but only three minutes later the fog mysteriously evaporated, and we were over Miami. How could that be possible when we had been close to 90 miles away and never saw the blue sky again until we reached Miami? We never even saw the chain of Bimini Islands that we had to fly over on our way to Miami.

It was thirty years later in the year 2000 and I am still researching the mystery and reading three different books that the famous author and pilot, Martin Caidin wrote about an experience he had flying near Bermuda. He explained in detail what happened to him when he encountered a strange fog that made all his electronic and magnetic instruments malfunction. He was in this fog all the way from Bermuda to the coast of Florida until he finally got out of it. I knew exactly what he was talking about because it must have been the same type of fog I had experienced.

When Caidin said the fog must have been a strange force field and he felt like he was flying inside a milk bottle, that is when that same light bulb went off in my mind again. That is when I realized the fog must have been attached to him and he never realized it, but he felt it surrounding the aircraft. In other words, the milk bottle was attached to him. Therefore, the fog must have been attached to me during that three-minute flight. That is why I never saw Bimini. That is why when I got out of the fog when I looked back expecting to see it-- there was no fog. We were flying in clear sky the whole time, but we could not see it because the fog made everything look obscure.

Years earlier I had given this fog the name "electronic fog." The term is not yet recognized by mainstream science yet, but it is known all over the world and I have been given credit for coining the term. I believe electronic fog is the main cause of many disappearances in the Bermuda Triangle. All pilots need to understand that if they encounter electronic fog, they are not flying through this fog but flying with it. If they do not realize this there is a high possibility, they may become spatially disoriented and crash. Unfortunately, most pilots do not know this.

TIMESTORMS AND THE TIME TUNNEL VORTEX

I call a storm a Timestorm if it produces a vortex tunnel. I have seen many of these tunnels and photographed many of them, most of them over the Everglades. Just recently my granddaughter, Kali Burton, helped me to capture what I believe is the first ever video of a time tunnel vortex. It formed on the southern edge of the Everglades near Florida Bay. These tunnels form when there are two cumulonimbus thunderstorms that are building up and are side by side of each other. Within each storm is what is known as a cell that is shaped like a vertical cylinder. When the cells build upward the storms usually connect on the bottom first and then anvil-heads form on the top and connect to each other creating a horizontal tunnel in between them. These tunnels usually last only a few minutes until they get so narrow, they collapse.

My theory is that when the tunnel collapses it emits a sphere of electronic fog. Like other fogs the sphere can last for many hours and depending on whether there are updrafts or downdrafts it can drift upward or down to Earth. If an airplane or person travels through it, it can instantly attach itself to them. I have talked to many people that believe they have experienced electronic fog—mostly in airplanes but some in boats, cars or on foot. It is the most dangerous for airplanes because they are unforgiving, and the pilot can become disoriented and crash.

What makes this fog completely different than any other is that it can affect the fabric of time. There are people like me that have been in it and gone forward in time or ahead of time. There are also people that have had time stop and there may even be people that have gone back in time.

I thought I had experienced the space-time warp while I was in the electronic fog, but a brilliant professor named David Pares from the University of Nebraska worked with me on this and has come up with another theory that could be plausible. He believes the tunnel I flew through was like what is known as a worm hole or a black hole. He thinks that when I came out of the tunnel, I was somehow almost instantly transported 90 miles forward in space to just ten miles offshore of Miami. I did feel that unexplainable feeling when I exited. After I arrived offshore of Miami it took me the normal three minutes to fly ten miles and reach the shoreline where the fog electronically detached. I had been instantly teleported through space.

THE POWER OF THE ELECTRONIC FOG

Over the past 200 years many distinguished scientists have theorized how it could be possible to travel through time. All have been criticized for being incorrect and not possible. This argument has been going on so long that anyone that says it is possible to travel through time is scoffed at. It is better to deal with linear displacements not time warps. None of all these great renowned scientists have ever experienced anything like what I have when it comes to space-time warps. We are worlds apart in how we think and feel about this subject. Their theories are all mental and my theories all come from and are generated from my physical linear displacement experiences with space-time warps. They may have witnessed subatomic particles and microscopic evidence, but none have had physical experiences like me. None of the scientists have ever ventured into the impossible like me.   My friend Dr. Valentine's famous quote indicates why I have discovered my theories: "The only way to discover the limits of the possible is to go beyond them into the impossible." Since I am the only living person that has flown through a time tunnel vortex, I feel compelled to continue researching and telling people about my experience until my end, especially to the younger generations because it will take a long time to solve this mystery.

The storm I flew through that was over the Great Bahamas Bank was like the mother of all timestorms. I have been studying radar almost daily for over 50 years and have never seen another storm that did anything similar to what that one did. But maybe a storm like that does not show up on radar. One popular theory is that methane gas could be causing this disturbance. The storm I flew through did appear to look natural and it was lying on the surface so maybe the methane gas was bursting out of the Earth and into the atmosphere and creating a giant Timestorm. Perhaps it was a giant bubble of this gas bursting out of the Earth and that is why it was in the form of a perfectly round circle—like a smoke ring.

There is one other theory that is possible, especially now that our government has released recently captured videos for the first time of what they prefer to call "unidentified aerial phenomena." Now we all know for sure because of these three videos that UFOs exist, but we do not know where they are coming from. Could these small UFOs be coming from another planet and are transported inside a much larger spaceship? Could it be possible that inside the strange lenticular shaped cloud I flew over, there was a huge mother ship? There are many people that claim to have seen a huge "cigar shaped" UFO hiding inside of what looks like a cloud.

Could it be conceivable that when I flew over the Lenticular cloud there was a mother ship inside of it and it was preparing to depart planet Earth to go to another star? Just as I passed over it, its propulsion system ignited, and I got caught up in it. The alien ship had harnessed the power of electronic fog giving them the capability to create a warp drive that has the ability to alter space-time. Our jumbo jets create a turbulence called wingtip vortices that can be dangerous if smaller airplanes enter them. This could be like our airplanes getting too close to the turbulence that a jumbo jet would create.

CHAPTER 8: MODERN HISTORY OF THE BERMUDA TRIANGLE

Strange events have been documented in the Bermuda Triangle ever since Christopher Columbus discovered the Bahamas Islands in 1492. He reported in his logbook that his compass was acting strange, and he saw a strange light in the night that looked like a candle bobbing up and down. That was so long ago we will never know exactly what he experienced. It could have been a bond fire from the top of one of the hills on the island that the Indian natives had lit, and the compass could have been slightly off because he was so far from his home his compass was off because of magnetic deviation. The magnetic North Pole causes this as you move around the globe.

The first mysterious event in the Bermuda Triangle took place in 1945 and I will tell you about that next after I tell you about the first well documented event that took place in 1928 but was not disclosed until 1978. The name of the pilot that had this experience was Charles Lindbergh. He became famous for making the first solo transatlantic flight in his plane, named the "Spirit of St. Louis," on May 20--21, 1927. He was only 25 years old and had only been flying for four years when he made the flight from New York to Paris.

However, despite his celebrity status—or maybe because of it—Lindbergh remained silent about a flight he made nine months later when he encountered conditions that puzzled him for the rest of his life. The story finally became known four years after his death, when his autobiography was published in 1978.

Lindbergh took off at 1:35 a.m. of Feb. 13, 1928, on the last leg of an around-the-Gulf-Caribbean tour. He would fly from Havana to St. Louis in what, for him, should have been a long, but routine flight. It would also be the first-ever direct flight between the two cities. "It should've been an easy flight—about a third the distance from New York to Paris," he wrote in his 1978 book AUTOBIOGRAPHY of VALUES. However, that is not what happened.

After take-off he climbed to an altitude of 4,000 feet and settled back to enjoy the night flight. "But halfway across the Straits of Florida my magnetic compass started rotating, and the earth-inductor-compass needle jumped back and forth erratically. By that time, a haze had formed, screening off horizons." Does this sound familiar? He is only about thirty miles south from where I had my experience with electronic fog and just like what happened to me his compass is spinning, and his electronic navigational equipment is malfunctioning.

Only one other time had he seen two compasses fail at once. That was during a storm in the Atlantic in route to Paris, and his magnetic compass only oscillated back and forth, so he was able to calculate his direction by the central point of the oscillation. But this time, near the Bimini chain of islands, the magnetic compass spun in circles (just like mine did) and the inductor compass was useless. "I had no idea whether I was flying north, south, east, or west."

Lindbergh started climbing higher thinking he could get above the fog and find the star Polaris; he could navigate by the stars. But the fog thickened as he climbed higher. So, he descended to less than a thousand feet, but the fog followed him, and he could barely see the ocean. Like all pilots at that time, he had no idea the fog was attached to him.

Just before dawn, he spotted a shadowy island and assumed that he had reached the Florida Keys. But after crossing a narrow body of water, Lindbergh saw a long coastline bending to the right, the opposite way that the land curved on his map of Florida. "But if I were not flying over a Florida key, where could I be? Was it possible I had returned to Cuba, that my attempt to read the twirling compasses had put me one-hundred-eighty degrees off course?"

He continued his heading, and he started seeing more islands ahead. He realized that if he was not over the Florida Keys, he was above the Bahamas. I believe that he had been flying in a northeast heading instead of the northwest heading he wanted to be on. He was about 200 miles off course and near the Abaco Islands. Once the sun was high enough above the horizon, he determined east and headed through the fog toward the sunny side of the fog. He spotted a coastline that was probably Great Bahamas Island and followed it until he reached the Gulfstream and then crossed over it to Florida. Just like what happened to me—when he reached the Florida coastline the fog electronically dissipated and at the same time his compass started working along with his electronic navigational instruments. From there he went on to complete his flight to St. Louis.

Possibly because Lindbergh was a renowned pilot, he never talked publicly about his strange experience in what would later become known as the "Bermuda Triangle." He waited to reveal it in his autobiography, which was published four years after his death in 1978. But no doubt he survived the experience because of his incredible abilities as a pilot, while many others have died.

Lindbergh's experience would be merely an interesting footnote to his flying career and amazing life, and nothing more, were it not for the fact that he documented a case of an aeronautical encounter with a rare, but often deadly meteorological phenomenon that remains a scientific anomaly.

CHAPTER 9: THE LOST PATROL

The most famous aviation mystery in history that took place in the Bermuda Triangle was in 1945 and it still plays a major role in the legend today. They call it the "Lost Squadron." It originated out of the U.S. Navy Base that is now Ft. Lauderdale International Airport. It involved five Navy torpedo bombers called TBM Avengers. They were labeled as Flight 19 because they were the nineteenth squadron on that day to depart on a training mission into the Bahamas and back. I am not going to go into all the details of their flight because there have been so many books written on the subject. I am only going to tell you what I think happened.

The squadron took off at 2:10 p.m. from Ft. Lauderdale and headed east while flying in formation to an area of deserted islands just north of Bimini called Hen and Chickens Rocks. When they arrived over these islands, they went into their routine practice of dive bombing for about an hour. The lead pilot, Commander Charles Taylor, was coordinating the routine. He was the most experienced pilot with over 2,000 hours of flight time. Just like me, the other four pilots had around 600 hours of flight time and they been all around the same age as me when I had my experience with the Bermuda Triangle. Each plane had two extra crew members except one member was missing so that made a total of 14 members in the squadron.

The practice routine of dropping bombs near Bimini was finished, they regrouped back in formation and headed east toward their next checkpoint. This is when I believe they encountered electronic fog. This is the first and only time that airplanes flying in formation have encountered this fog. At this point they are only about 30 miles to the north of where I was when I penetrated the inside edge of the timestorm in 1970. They are not only near the same location, but it is nearly the same time of the day, and it is not only the same time it is exactly 25 years later plus one day when I had my experience. Is this just a coincidence or is there a time correlation as to how the weather or scheduled timetables can repeat themselves?

At 3:40 p.m. Taylor made his first distress call on the radio and admitted he was lost. "I don't know where we are, we must have got lost after that last turn." This is nearly the same time that I made my first distress call, only I did not admit I was lost. At this point I believe the electronic fog had attached itself to all five airplanes and they all think it is just an area of haze. Disorientation has already set in, and it is impossible for them to determine what direction they are heading because their instruments are doing the same thing as mine were--malfunctioning.

Lieutenant Michael Cox, a senior instructor at the naval air station, who was flying over South Florida at the time, overheard Taylor and was able to contact him on the radio and asked what was wrong. Taylor replied: "Both my compasses are out and I'm trying to find Ft. Lauderdale. I am over land, but it is broken. I'm sure I'm in the Keys, but I don't know how far down, and I don't know how to get to Ft. Lauderdale."

Now both of Taylor's compasses are spinning and his electronic direction finder is malfunctioning. He has no way to navigate except by using his mind and it appears his mind is also disoriented. Electronic fog can induce spatial disorientation, but these pilots were highly trained and young and able to avoid it. Taylor is so confused he thinks he is over the Florida Keys—over 100 miles off course.

Cox asked Taylor: "What's your present altitude? I'll fly down and meet you." After a moment, Taylor replied. "I know where I am now, I'm at 2,300 feet. Don't come after me." Did he say this because he knew something supernatural was happening and he did not want to endanger anyone else? I know when I was in the electronic fog, I would not want anyone flying close to me because I was flying in something I did not understand and felt as if it could be dangerous for anyone to try and locate us.

It appears everyone in the squadron was confused and disoriented and are having the same electromagnetic problems. When you are caught in this fog it is best to immediately make only one turn toward a safe direction and in this case, it would have been to the west back to Florida. The more turns you make the more it becomes impossible to determine what direction you are flying. It appears they went through a series of up to seven turns before they realized they were running out of fuel. They were overheard on the radio preparing to ditch the planes all together since they were over the ocean with no land in sight. Minutes later they were never heard from again.

One of the largest rescue missions took place in the search for them but nothing was ever found. It bothers me that not even a life jacket or a piece of the planes were ever found. They were all trained in how to ditch, Taylor had already done it twice before. They were all wearing their life jackets and they also had life rafts on board, and they were trained in how to use them, but not one piece of evidence was ever found. Something should have eventually floated to a distant shore sooner or later. They were lost forever without a trace.

Landing an airplane in the ocean is hazardous, but not impossible. If the squadron had attempted the water landing, one or more of the crews should have been able to open the hatch and drop their self-inflating life raft into the water before their airplane sank. The life rafts would have been found by the rescue team. But they did not find anything. That means the planes may not have made the water landing they planned. Something else happened to them. But what? Is it just a twist of fate that nothing was ever found, or did they encounter a powerful electronic fog and are no longer visible on planet Earth?

Some transmissions have been forgotten because they were heard by amateur Ham radio operators and the only official recordings left are from the Navy and some of them have been left out. One of the forgotten transmissions long ignored came from the flight leader Taylor saying "We don't know which way is west. Everything is wrong…Strange…We can't be sure of any direction—even the ocean doesn't look as it should." This sounds like the exact same situation I was in when I was inside the electronic fog.

Another comment from an unidentified pilot said: "Every gyro and magnetic compass in all the planes are going crazy." Some other comments came from Marine pilot Captain George Stivers who had taken over command apparently because Taylor had become incompetent and gave him the command. He said in a panicked voice "It looks like they're from outer space! Don't come after me!" Could he have seen a group unidentified aerial phenomenon?

A board member of the Coast Guard that requested anonymity was quoted in the local papers in 1945 as saying "They vanished as completely as if they had flown to Mars." Also, in the Miami News was a quote from my friend and scientist that lived in Miami that I have done research with on the Bermuda Triangle, Dr. Manson Valentine, saying, "They are still here, but in a different dimension of a magnetic phenomenon that could have been set up by a UFO." Dr. Valentine and I have something in common. When he passed away at the age of 92 in the year 1994, he was known to have been researching the Bermuda Triangle longer than any living person-- for almost 50 years. I have also been recognized as being known to have been researching the Bermuda Triangle for longer than any other person-- for just over 50 years.

Dr. Valentine and I contacted each other in 1976. He was one of the smartest people I have ever known, having three doctor degrees from Yale University. We worked together on the mystery and in 1977 he co-authored a book with best-selling author Charles Berlitz called WITHOUT A TRACE. He had me play a major role in the book and to end the book he made some comments that I believe are still true to this day. "Either the magnetic fields are the result of sporadic, perhaps seasonal buildup of geophysical origin, or they are the concomitant effects of UFO activity. Possibly a combination of both agencies can occur.

Valentine was also familiar with the time tunnel vortex. When I met him, he was probably the only person that had done research on the tunnel phenomenon. When I explained to him what the inside of the tunnel looked like he was fascinated to be able to talk to me about it because he realized I was the only living person with knowledge of what it looked like. He went on to say in his book something important about me. "Like the tornado, the magnetic vortex would be self-augmenting and could well bring about an inter-dimensional transition, for anyone caught up in it. The experience of Bruce Gernon is a case in point." It meant a great deal to me that a renowned scientist made a public statement about me that my experience was the "case in point." He also told me to "Never forget that you hold the key to the mystery."

CHAPTER 10: THE GREATEST MODERN AVIATION MYSTERY

The most people that ever disappeared in an aviation mystery took place in Malaysia on March 8, 2014. The flight MH370 departed Kuala Lumpur Airport at 12:41 a.m. local time; aboard were 227 passengers and 12 crew; the captain was Zaharie Ahmad Shah. The airliner was a Boeing 777-200, an airplane that has one of the best safety records in commercial aviation. At 1:01 a.m. they reached a cruise altitude of 35,000 feet and were near the north coast of the peninsular of Malaysia. At this point they accelerated to cruise speed, and everything seemed normal.

At 1:07 a.m. the last active transmission from the aircraft's automatic electronic reporting system is received. This reporting system is timed and uploaded to a satellite. At 1:19 a.m. Malaysian Air Traffic Control tells Flight 370 to contact Vietnam at 128.4 (radio frequency). This is a common procedure when preparing to enter another area of controlled airspace. The co-pilot answers and says: "All right goodnight." This is a rather casual way of signing off, so everything still seems to be going well. That was the last transmission ever heard from Flight 370, although another airliner that was flying nearby said they heard them talking on the radio a few minutes after that last call, the transmission was garbled and had static. I believe there is a good possibility that the mysterious disappearance of the Malaysian Airlines was caused by electronic fog, and it was at this point in time when the attachment was taking place. The airliner had just entered a sphere of this fog and it attached itself to the aircraft. Therefore, the electronic instruments and the radios are shutting down. At 1:21 a.m. the MH370's transponder stops transmitting the regular responses it is programmed to give to radar, identifying the plane's altitude, speed and bearing.

Perhaps the electronic fog disabled the radios, and all the readings on the Boeing 777's glass panel cockpit turned blank. The pilots had no idea of their exact heading because even the wet magnetic compass was spinning. At that point, they were relying on mechanical backup instruments—the altimeter, the airspeed indicator, and the attitude indicator—to maintain control.

If the pilots had known about electronic fog and realized it had attached to them, they could have used my advice on how to escape it. Since they were over a large body of water when it attached, they should have maintained their heading and not make any turns. If they had done this, they would have eventually reached the large body of land of Viet Nam and the fog would have electronically detached. But instead, they made their first fatal mistake and made a 120 degree turn to the left, probably trying to aim for the nearest airport.

Confusion would be setting in. They tried something that other pilots have attempted, like Lindbergh and Martin Caidin, but they were unsuccessful because they did not realize the fog is attached to the airplane. They violate regulations and climb up trying to get over it. Since they realize this is an emergency, they go above the maximum altitude allowed for the Boeing 777—all the way up to 45,000 feet and they are still in the fog. Then they try to go under it. They descend all the way down to 20,000 feet and they are still in the fog.

At 1:30 a.m. Malaysian radar detects the airplane for a final time, and they have made another turn to the northeast. At 2:25 a.m. Malaysian military radar detects them and now they have made another turn to the northwest. That was their last radar contact. After another hour, a special U.S. government satellite makes contact and sees the airplane making a fourth turn to the south and even making circles. After making four turns and then several more circles they must be spatially disoriented, and the autopilot is maintaining some random course. They have lost their sense of direction and if they are still conscious, they would have no ideas left as to what to do next.

Maximum flight time because of fuel exhaustion will expire at 8:40 a.m. Their situation is like that of Flight 19's. They would continue flying for four more hours, not knowing where to turn to, until their fuel ran out and then made a crash landing into the ocean. Also, like Flight 19 they disappeared in a remote location of the ocean where they may never be found.

An extensive search was made for them by air and there was no evidence found anywhere but the area was so extensive that was to be expected. A search by water was made that lasted over a year by ships using the latest submersible technology but nothing was found. After a few years over thirty pieces of debris from the airliner washed up along the African coast and a large piece of the wing called the flaperon washed up in Madagascar.

The debris proved to be positive evidence that they were part of Flight MH370. This also indicates the airliner did not disappear and did not encounter the strongest type of electronic fog. Engineers said that the way in which the flaperon was damaged indicated that the airliner crashed without using its flaps to slow down the plane before it landed in the water. This indicates the autopilot was probably flying the plane until it made the crash landing in the ocean. Boeing did a simulated animation of what the 777 would do if it were on autopilot and it ran out of fuel. The starboard engine would turn off first and the autopilot would engage full left rudder to correct the right yaw. Then the port engine would shut off and the autopilot would disengage. If someone were flying the airplane they could glide forward and engage full flaps before making a controlled emergency landing in the ocean. If no one were flying the airplane it would enter a large spiral and make a hard-crash landing and nosedive into the ocean. This appears to be what happened because the flaps were never engaged, and the airplane broke into many pieces. This indicates that the pilots must have passed out or everyone was already dead upon crashing. Electronic fog has been known to put people to sleep.

When the Malaysian airliner's radio and transponder went off the captain could have turned them off, but he could not have turned off several other electronic reporting devices that were malfunctioning, it would have taken an airline mechanic or perhaps electromagnetic interference like electronic fog.

The Malaysian officials say: "There are no new clues and there is no conclusion as to why the plane vanished." On July 19, 2019, a 19-member international team said: "There was no evidence of abnormal behavior or stress among the two pilots that could lead them to hijack the plane."

With no recognition of electronic fog by mainstream science, my scenario is not even being considered as a possible solution to the mystery of the Malaysian airliner. Electronic fog is a rare phenomenon, and I know it is real not only because I have experienced it, but so have many other pilots and once it becomes recognized lives will be saved.

Scientists have discovered that there are rogue waves that seem to come out of nowhere and are capable of sinking boats and ships. This is the most likely cause of the disappearance of many of the boats in the Bermuda Triangle. They have also theorized that methane gas has sunk ships, but this would be exceedingly rare. Whether this gas could affect airplanes I do not think so. But I must admit that the timestorm I flew through did look like it was a natural phenomenon and the way it was laying on the surface of the ocean made it look as if it could have come out of the Earth. Could it possibly have been a giant bubble of methane gas that burst out of our planet and ignited the atmosphere? If this were true it would be possible to affect airplanes also.

CHAPTER 11: THE ALIENS

The term "UFO" was coined in 1953 by the United States Air Force (USAF) to serve as a catch-all for all such reports their pilots could not identify. It stands for unidentified flying objects. It means things in the sky that were not identifiable at the time of observation. Many believe the ones that remain unidentified are extraterrestrial alien spacecraft. Today the Air Force continues to investigate UFO's and has changed their name to "UAP" which means unidentified aerial phenomenon. This makes sence because we still don't know exactly what they are. Are they natural creatures from our own planet Earth, our solar system, our galaxy or the universe? Are they controlled by aliens from another world? Throughout history they have been known to be controlled by what our ancient ancestors call the "star people." In recent times the Navy fighter jet aviators have captured them on video in great detail with their sophisticated infrared radar because they have been violating our controlled airspace without permission. The videos show them doing maneuvers that we are not capable of performing. Even with these new up-close images we still don't know for sure what they are.

THE FIRST UFO

When I was ten years old, I moved into a new home with my mom and dad in Boynton Beach, Florida in 1956. One year later my cousins moved into a new home right next to us. They had a family of seven. One night shortly after they moved in, they returned home after watching the movies at a drive-in theater around 10:00 p.m. After they parked, they ran out to the street between us and yelled out to us, "Come out of your house you guys, you've got to see this!"

We hurried outside to see what all the commotion was about. They said while they were driving back home in their convertible car, they spotted this strange light in the sky that appeared to be going in the same direction as them. They said it had been making a lot of strange movements. They pointed it out to us, and I watched it moving slowly from the right of us and then it stopped and hovered right even with us to the west over an area of undeveloped land on the edge of our subdivision.

It was about 1,000 feet from us and about 200 feet high. It was a perfectly round sphere about 35 feet in diameter. It was glowing bright white in color, and it appeared to be in the form of a solid state, not like a cloud or gas. It was bright like a florescent light but appeared to be metallic. It made no noise as it sat there hovering for a few minutes. All ten of us watched it in amazement and had no idea what it could be. Then it started oozing out some strange streams of what appeared to look like contrails. Some researchers speculate these white streams may have been plasma flowing outward just before takeoff. They started emitting from the sides and bottom of the object. They were the same color as it and then slowly streamed downward swirling and then dissipating after reaching a length of about 50 feet. This lasted for about one minute. I have seen an identical photo showing the spere with the streaming plasma, someone captured in a book that was written about the Bermuda Triangle in the late 70's. We all watched in awe as this incredible event took place and then after another minute it suddenly took off at an incredible speed and went straight up and out of site within only a couple seconds.

Lately there have been a lot of videos of what are called orbs on TV that were caught on camera by people with their cell phones. Some appear to be type of ball lightning like the five I photo'd in the late 1970's in formation inside a thunderstorm over the Gulfstream late at night. Many of the videos have shown multiple orbs moving in strange formations. Others appear to be like the one I saw in 1957. The one we experienced may not be a natural phenomenon. It seems like it was under some type of control by a higher intelligence, or it was alive and has its own intelligence. Perhaps it was under remote control.

(This photo was taken in 1975. The five orbs appear to be in formation inside a thunderstorm in the same area of the Bermuda Triangle where the famous Flight 19's five bombers disappeared while flying in formation.)

CLOSE ENCOUNTER of the FIRST KIND

It was only one month after I had my experience in the Bermuda Triangle, I had a "close encounter of the first kind." I was living alone in a condominium on the ocean, and it was a perfectly clear night. The wind was light, and it was CAVU—clear air visibility unlimited. I ran into a neighbor girlfriend and asked her if she would like to go flying with me. She had never flown at night in a small plane and decided she wanted to go with me. We drove together to my airplane at Palm Beach International and lifted off around 9:00 P.M. We headed due south along the coastline passing by our apartment and headed toward Miami.

We could clearly see all the city lights all along the "Gold Coast." I kept climbing to get a better view. The lights became more concentrated as we approached the City of Miami. We went all the way to 10,000 feet before leveling off and accelerating to a slow cruise speed. When we were directly over Miami International Airport, we headed due east while enjoying the spectacular maze of lights below. We crossed over Miami Beach and continued east over the Atlantic Ocean as we left the city lights behind. When we passed over the beach, we left all the lights behind and everything in front of us was pitch black—there was no moon out—all we could see was the stars above and the black ocean below us. When we were a few miles offshore, the darkness of the sea and sky seemed to blend together looking like a vast black abyss.

I was in the same area where I had exited the electronic fog, when I noticed an orange light far off to the east that appeared around the size of a planet. It was just above the horizon and seemed to be moving slowly. It was making odd movements, going up and down and side to side and bobbing around. I asked my girlfriend if she saw what I saw, and she agreed it was strange. Suddenly, the orange light grew larger, as we watched it in amazement. It was moving directly toward us at an incredible speed and appeared disk-shaped.

Within ten seconds it was right in front of us, and it was enormous. It was disked shaped and appeared to be over 100 yards wide 30 yards thick. It had a dome on the top side of it. It was bright amber and it looked metallic. It was about three times the mass of a Jumbo Jet. I thought we were going to be demolished. It came so close that it filled the entire windshield.

Just before impact, I veered sharply to the left, turning as hard as possible. I thought we had no chance of avoiding a collision since we were so close to the object. I estimate we were within 50 feet of it.

Somehow, we avoided the collision and after I made the turn, I looked back expecting to see the UFO moving in a westward direction. To my surprise, it was gone as if it had instantly vanished. I concede that it could have lifted upward and soared skyward like the one I watched when I was ten. We were just relieved to be alive. We were both shaken by the experience and maintained our northerly heading and descended back to Palm Beach Airport without further incident.

The UFO was shaped like a classic flying saucer. It was domed on the top side and flatter on the bottom. It had another dome on the top side that could have been the cockpit area. There were no visible windows and the entire UAP appeared to be glowing bright orange. It was such an amazing experience I wondered if it might have been an illusion, but my friend saw the same exact thing also. So, what was it? I did not think it was an alien spacecraft because back then I did not believe in them at that time. I thought it might have been a form of lightning that had not been identified yet. But my belief of UFO's has changed over time.

    This experience with the UFO and the Bermuda Triangle seems to have something to do with fate and I even have a hard time believing that it really happened. Why would I climb up so high to 10,000 feet? Why would I go offshore and head toward Bimini? By putting the airplane in this position, it put us on the same altitude and flight path that I took when we exited the time tunnel vortex and ended up over Miami Beach, only we were traveling in the opposite direction. With these two events taking place in such a short period of time, it seems like someone was trying to tell me something. These were major events in my life, and I took them in stride thinking that it had not changed my life—yet little did I know.

STUDYING TIME AND UFO'S

After I learned that I was part of this mysterious area now known as the Bermuda Triangle I became obsessed in researching everything I could that had to do with time and UFO's. I was excited about it because I thought I had discovered through my experience the cause of what creates the mystery. I would spend the next three years going to the library usually three times a week and spending a few hours reading every book available about time and UFO's. I must have read about 200 books during those years. When I look back in my memory there was one book that stands out above all the others that I had read in the library, and I cannot remember the name of it. It was written by a South American scientist that was a doctor and it was published around the fifties or sixties because I remember that the book like it was old and warn back in 1972.

The book was all about UFOs in South America and it featured photographs the Doctor had taken using a telescope. The most intriguing photos were of what could be called a Mothership. He captured the photos of it orbiting high above the earth. It appeared to be exceptionally large, maybe close to a mile long and one quarter mile wide. It had the look of a famous UFOs that had been seen before for centuries and is known as the Cigar shaped UFO. There were clear photos of it with several different magnifications. As the photos zoomed in you could see an unusual feature of the exterior of the ship. It had strange, shaped dimples on its skin. They were kind of like the dimples on a golf ball, but they were all different sizes. Perhaps that's why they call it cigar shaped. It not only looks like a cigar, but it is also not a perfect symmetrical shape, it could be called asymmetrical.

The most amazing photos the Doctor captured were the close-up photos of the side of the ship. There was an opening on the side, and you could see a square hallway leading into the ship. There was a saucer shaped UFO just outside the hallway and it was either coming out or going into the opening. The flying saucer looked identical to the one I encountered offshore of Miami Beach in January 1972. When I saw these photos, I did not know what to think. Were they real or fake? Were they of something from nature or were they man made? Whatever he captured has left a strong impression on my mind even after 48 years.

REMOTE VEWING

Joe McMonagle, the famous "remote viewer," recently told me something I can relate to the Doctor's photos. Joe used his remote-viewing talents and has seen this strange Mother ship in his mind. He said it is alive! It is a living creature that can travel throughout the Universe and the Aliens have learned to use it and live inside it? Joe said he tried to get inside of it so he could see how they controlled it, but they have only let him briefly see it a few times. Joe said, "In those instances I've seen control systems which appear to be part biologic and part physical, so I would have to say their ships are hybrid systems made up of grown materials within material frameworks. They communicate with their ships much as we would consider mind-to-mind communications to be." That reminds me of how early man's relationship with horses could be analogous to the Aliens relationship with this creature. At first man used to use the horses as a food source. Then they realized there was a special relationship with them. They stopped eating them and learned to use them for travel. This was a giant step for mankind as the "time" to get from point A to B was drastically reduced by making friends with the horses and learning how to ride on them.

McMonagle also talked about how aliens are sensitive to our viruses. This is a coincidental situation today now that our planet Earth is confronted with one of our worst pandemics, the Corona Virus. He said, "I've also seen what I call their "skin Suits" which are the environmental suits they must wear if they are exposed to our atmosphere. They are highly vulnerable to our biological systems. They are deeply afraid of the many viruses and biological agents we carry within our bodies, and which are swimming in our air and waters or found within our animals and plants, and the very dirt at our feet. They have little immunity to them. So, they wear hazard suits, or what I call skin suits to protect themselves.

They also protect them from the agents they must immerse themselves in, to eradicate these elements whenever they re-board their ships. These skin suits hide their features completely and make them all look the same. The large eyes people report are the protective lenses that cover their eyes." I wonder how they protect themselves from the airborne germs in our atmosphere. Their mouths may be covered by some sort of mask but there does not appear to be anything covering their air openings to breathe.

"Likewise, these skin suits protect them from us as well. Our natural reactions would kick in if one of them appeared in their natural state to us. Our reactions would be instant and violent. They understand that we are a violent and reactive species and that is one of the reasons they do not contact us openly, but only do so when the circumstances are right, and they are in full control.

We think we are their equals, but this is simply arrogance. They are half a million years ahead of us, in capability. They jump star-to-star without effort and operate on ancient rules that far and away transcend our understanding of how things work. Our belief in their abilities as alien creatures is pitifully under reaching, and we are quite primitive in our understanding for their limits and abilities. And while we might consider them butt ugly, they consider us to be half a step behind a chimp in our development."

I believe this cigar shaped mother ship could be real and might be the cause of what I experienced in the Bermuda Triangle. Inside the lenticular shaped cloud, I flew over, there may have been this spaceship. I have been studying radar images in this area for over 50 years and have never seen a storm formation quite like the Timestorm I encountered. I have seen several formations that are similar, but they took several hours to form, not 20 minutes like the one I flew through. I have seen many tunnels between thunderstorm cells that were like the one I flew through, and they formed naturally. The one I flew through in 1970 did appear to be natural although it may have been created from the exhaust of the mother ship. It could be something like the dangerous wing tip vortices that are created from the large jumbo jets. The turbulence I flew through may have been the vortices created from the cigar shaped Mother ship that just happened to be embarking for another star right when I flew over it.

(Formation of thunderstorm vortex tunnels. The second photo taken one minute later shows a tunnel about to form within only a few seconds, on the left side of the storm. The photos show a previously formed tunnel on the right side that collapsed about one minute later)

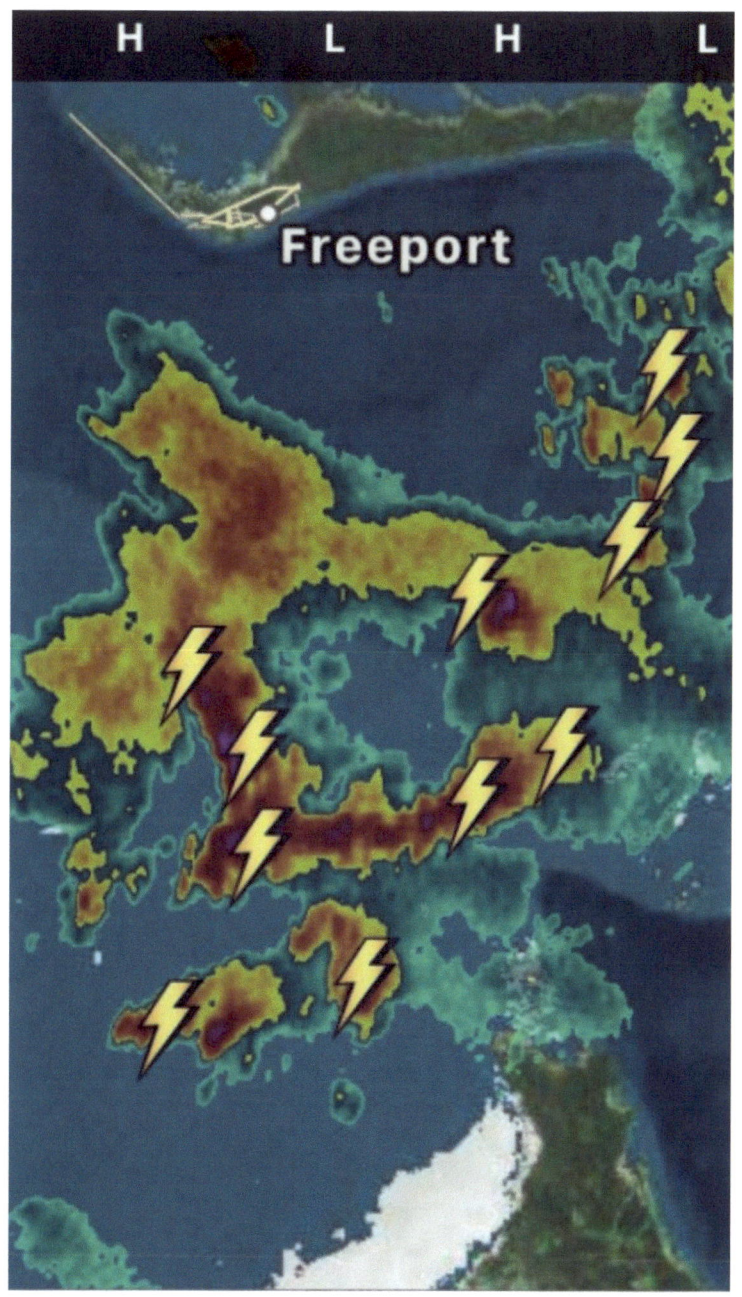

(Radar image of a circular squall that has formed between Grand Bahamas Island and Andros Island. Note on the right side there is an opening that could connect and form a vortex tunnel)

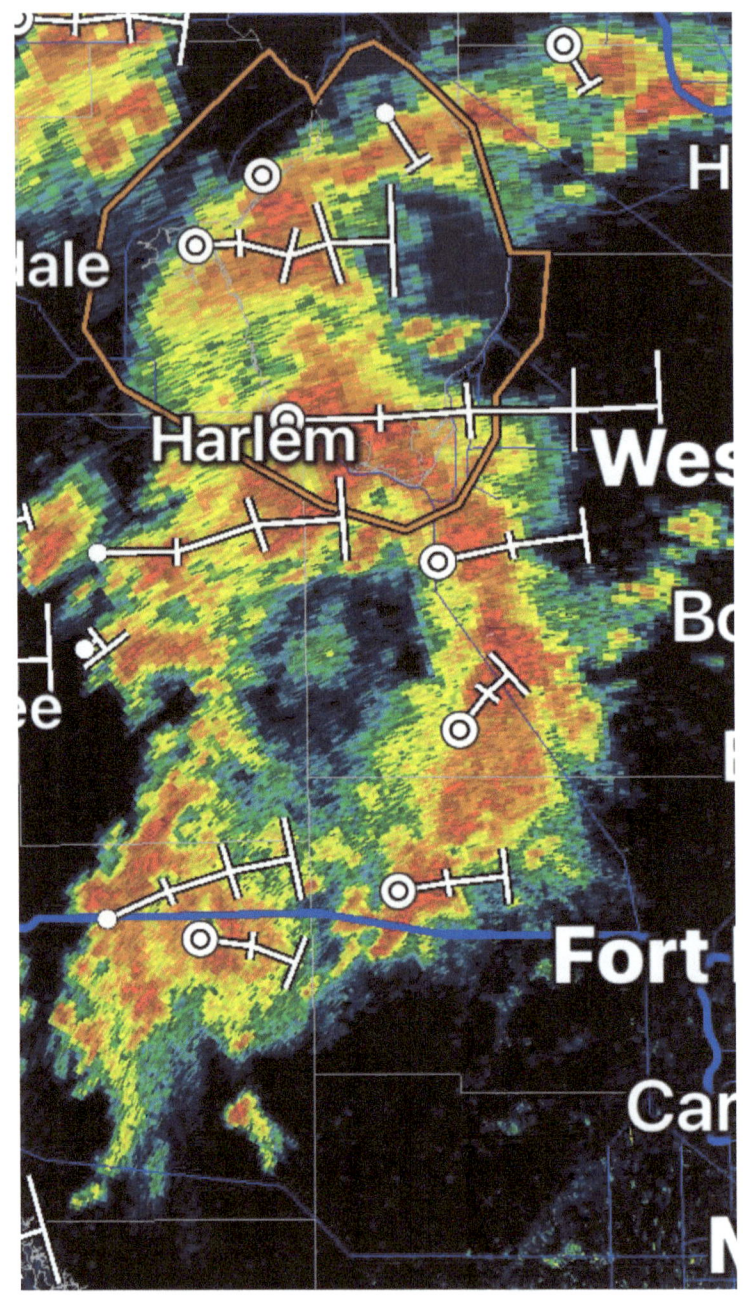

(Radar image of a circular squall over the Florida Everglades. This storm is so intense that if an airplane were caught in the center of it, the only way out might be through a vortex tunnel)

One of the most unusual timestorms I've ever seen formed over the southern edge of the Everglades. My granddaughter Kale Burton was with me, and we watched it form from its birth to maturity as the sun was setting in Key Largo. As the storm reached maturity it formed a vortex tunnel. Kali was smart enough to capture this event on her cellphone. I believe this is the first and only thunderstorm timestorm with a vortex tunnel to be taken on video. The tunnel was about five miles long and it lasted for about four minutes. At one point a lightning bolt went off on the far-right side of the storm and it lit up the tunnel on the left side of the storm. Just before the tunnel collapsed it did something unusual—it appeared to form the shape of a square rectangle like a hallway that only lasted for a few seconds. There have been hexagon shaped clouds that scientist have reported in the Bermuda Triangle that appear to be a part of the mystery.

(Timestorm over the Everglades near the edge of Florida Bay. A lightning bolt went off inside the storm on the right side and it lit up the vortex tunnel on the left side just after sunset)

(Just before the vortex tunnel collapsed something unusual happened. The tunnel formed the shape of a square rectangle. It appeared to look like a hallway!)

(My Granddaughter Kali Burton and me on the fourth floor of "Jimmy Johnson's FISHERMANS COVE resort, with the timestorm in the background)

UFO PRECOGNITIONS

In 1974 my wife Lynn and I flew in a commercial jet airliner from Palm Beach to North Carolina for a snow skiing vacation. While we were planning this trip, I kept getting the idea that we would see a UFO while flying back to Palm Beach. I had no idea why I kept thinking this, but I felt confident it would happen. I told Lynn about it, and she did not believe it would happen. We booked the two seats for the flight back to be on the left side of the plane because I knew the UFO would be somewhere just off the Florida shoreline.

We had a great time skiing, and I was looking forward to the flight back. I was starting to get excited when we boarded the plane and got our seats by the window. I let Lynn have the window seat because she was better at taking photos with our camera. We started our decent into Palm Beach and were about 30 miles north and at 3000 feet when it happened. It was around 9:00 P.M. when we spotted the orange light going in and out of the clouds just offshore. It was near the same time I had encountered the flying saucer offshore of Miami in 1971.

It looked like the same UFO I had encountered offshore of Miami Beach. It was about one quarter mile off the shoreline and flying only about 100 M.P.H. and heading south and level at 2,000 feet. We were about two miles away from it. It was the same bright glowing color and the same size as the one I saw near Miami, and it looked like a solid mass. It also had the same dome on the top. There were scattered cumulus clouds with their bases at 1,500 and the tops at 3,000 feet. The craft was going in and out of the clouds and when it entered a cloud it would light up the entire cloud with a lite orange color. People on the other side of our aisle were starting to stand up to get closer out the windows in order to get a better view. We only got to see it for about a minute as we flew by it at around 200 M.P.H. When we got off the plane no one said anything about the UFO. I think everyone was in shock and did not know what to say. I shook hands with the captain, and I could tell he had seen it but back then we did not talk about things like that, especially airline pilots. Lynn took several photos of it, and they turned out, but the quality was poor. We were just using a standard lens. If we had been using a telephoto lens it would have turned out better, but you could see the orange color and in one of the photos you can see the dome on top of it.

FLYING IN FORMATION

The next day I started getting those same thoughts again that I would see another flying saucer. I was sure it would be around the same time again and it would be over the ocean. I told Lynn about it and this time she believed it might happen. We drove to the nearby beach at Delray Beach shortly after the sun went down. This time I brought my binoculars that had an electronic zoom.

The skies were clear, and it was just after 8:00 P.M. when we spotted a UFO about two miles offshore. We watched in amazement as it came into our vision from the far north. It was the same glowing orange color as the one we saw the night before. It was about 2,000 feet high and heading due south and moving fast, about 500 miles per hour. In less than ten seconds it was out of sight.

At the same exact time that it disappeared another identical UFO appeared to the north, and it was on the same identical flight path traveling at the same speed. We watched it fly by us and again as soon as it was out of sight another similar UFO appeared from the north on the same flight path. This time I grabbed my binoculars and focused on it. It appeared to be the same type of flying saucer configuration as the one we saw the night before. As soon as it was out of sight a fourth UFO came from the north on the same flight path. This time I zoomed in on it with the binoculars again and there was no doubt it was the same type with the dome on the top.

We were watching an incredible event and were excited, but the awe-inspiring grand finale was the last UFO. The fifth and final UFO appeared in the same fashion as the previous other four and it did something impossible right in front of us. When it was exactly even with us and directly to the east of us it made a turn toward us. But it was not a normal turn. It had no radius in the turn. The turn was a square ninety-degree turn. It instantly turned toward us within a fraction of a second while traveling nearly 500 miles per hour.

Within a few seconds, it was directly over our heads. It made no sound whatsoever as it passed over the shoreline and continued heading west. Just after it passed over us it emitted two bright flashing lights, the first was red and the second was green. We watched it for a few more seconds before it disappeared over the Everglades. My friend and co-author Rob MacGregor wrote an article about this event, and I often ponder about what he said about me after the flying saucer passed over my head, he said, "It was as if the UFO was saluting Gernon."

I wonder if that last UFO that made an incredible turn right when it got even with me and flew right toward me and then started flashing lights when it got directly over my head was trying to give me a message? I have the same thoughts about the UFO that almost ran into me offshore of Miami. Why did it travel on the same exact flight path that I had traveled on a month earlier when I went through the tunnel vortex and what coaxed me to go on that flight path on that night? Were they communicating with me and showing me their incredible abilities to maneuver their spacecraft?

Did I contact them? If they were trying to communicate with me, I do not understand the protocol.

Having studied airplanes and UFOs for seven decades I feel I have an attraction to them. If humans continue to survive for more centuries, it seems we will eventually be capable of travelling to other stars that will have planets like ours and there will be beings like us. It appears that aliens are already centuries ahead of us and are visiting us and have been doing it for thousands of years. I have appeared on two "Ancient Alien" TV documentaries about ten years apart and it looks as if their popular TV shows will be continuing for many more years. They are interested in my experience because they must believe it may have been caused by aliens. I believe they could be right. Could my flight have been influenced by aliens? Was it just an accident or were they trying to tell me something-- or could it have been both? I am glad I have never seen or met any aliens because they are rarely friendly to humans and if they do meet with you, they have complete control of you apparently by using the tremendous power of their minds. They may want to be friendly to us and help us, but it would be like us trying to help gorillas.

CLOSE ENCOUNTERS OF THE THIRD KIND

In 1994 perhaps the most significant close encounter of the third kind in modern history took place with aliens and they were friendly. It happened in the country of Zimbabwe located inland in

South Africa. There had been UFO sightings near the town of Ruwa for three nights in a row. Hundreds of people witnessed them, and some videos were captured showing a glowing orange disk flying overhead. This event took place in daylight on the fourth day right after all the sightings. It happened at the local schoolhouse called Ariel where first through seventh grade students attended. It was recess time, and all the 64 students were outside playing in the playground. Four silver oval shaped flying saucers were seen flying above them and one of the spaceships landed on the edge of the playground while the other three remained motionless hovering above. Most of the children went into a state of panic when they witnessed this. To everyone's amazement an alien being emerged out of the ship and walked right up to all the students. The creature was dressed in a black tight fitting jump suit. The alien communicated with the students telepathically while it walked within three feet of them. After doing this for a few minutes the alien went back inside the spaceship and departed along with the other three saucers and was never seen again.

    The school had students that were between the ages of five to twelve years old. Shortly after this incredible event took place and word spread about what happened, the children were videoed and interviewed. A renowned professor from Harvard University even came to question the children and six of them were videoed. Some of them said they feared the alien because they had "never seen a person like that before." Most of the children had no fear of the alien once it started to communicate with them. One little girl said, "It was about as big as a sixth grade (student)." They were all fascinated with the eyes of the alien. One boy said, "The eyes were four or five times bigger than ours." Another said, "Maybe they are trying to communicate with us and show us something we don't know about." One girl said, "I think they want people to know they actually don't want to harm this world." This type of alien is known as a "Gray" and has been reported for thousands of years. Many of the students drew sketches of the spaceships and the aliens and they looked like shapes that have been reported worldwide. There were similarities in all of the drawings…they all showed the spaceship had landed between two large trees, the spacecraft was long and silver with a series of many bright lights along the bottom, and a ramp with stairs extended from the craft to the ground along with an alien that had large, slanted eyes.

Twenty years later those same six students were interviewed again, and they all still struggle to understand what happened. They do not talk about it often. They all felt they knew they did not have to be afraid of the creature even when they had close contact. They all said they could not forget the mesmerizing eyes, they seemed to be hypnotic. Some of the comments the now thirty something year old students said was: "It was trying to communicate with us, maybe something to do with the environment." "It was trying to tell us technology is not helping, technology is bad." "We are going down a wrong path, we have to start recognizing what we are doing is detrimental and we need to make changes." "They were reaching out to us—it was as if they wanted them to go with them." What changes do you think we need to make that the alien was referring to?

These witnesses were asked how long of a time did this event take? They all paused for a moment before one of them said, "It was difficult to judge time." They all agreed. Then one of them said, "Maybe about twenty minutes, but it seemed much longer." Was time being altered when they had this encounter? Many people who claim to have had encounters of the third kind say they lost time. When they realize what time it is, it is much later than they thought. In this case it could be what is known as a time distortion or a temporal time dilation of compression. Because it was a life altering event, time seems to slow down, and they remember every possible detail for the rest of their life. This is what happened to me when I flew through the time tunnel vortex—it seemed like I was in the electronic fog for about thirty minutes, but it was only three minutes.

One of the teachers that was monitoring the children and witnessed the entire event was interviewed. She was by this time promoted to the principal of the school. She was ashamed of herself for what she did at the time—she was so scared she stayed in the background in shock. She was asked if she could put in words one sentence about what happened that day. She said, "We were visited by aliens."

Are we sharing this planet with an unknown other? One with a mysterious relationship to humanity and its own interests in our world? Why is this most incredible encounter has been with children and not adults? Could it be because the future of our planet is in their hands and the present adults are doing something "detrimental" to our world? Something like having continuous wars and threatening nuclear war?

Perhaps when I made that flight in 1970, the lenticular shaped cloud that just happened to be right in front of my flight path was hiding a huge mothership cloaked inside a cloud. It was just hovering there over one of the remotest areas of the Bahamas and preparing to depart planet Earth. As fate would have it, its countdown to take off went to zero just as I climbed over it, and I got caught inside its exhaust and departure turbulence. The aliens knew I was caught inside but it was too late for them to delay the launch. Fortunately, I was able to escape the turbulence after being caught up in it for twenty minutes. After finally escaping their turbulence one of them contacted me telepathically and told me how important it was to remember what I had just experienced. The exhaust from this phenomenal spaceship was an incredible force. This is why I call it a timestorm. They were not just departing like one of our rocket spaceships going to outer space—they were departing to another dimension.

REALITY CHAPTER 12: THE POWER OF THE ELECTRONIC FOG

Over the past 200 years many distinguished scientists have theorized how it could be possible to travel through time. All have been criticized for being incorrect and it is not possible. This argument has been going on so long that anyone that says it is possible to travel through time is scoffed at. It is better to deal with linear displacements not time warps. None of all these great renowned scientists have ever experienced anything like what I have when it comes to space-time warps. We are worlds apart in how we think and feel about this subject. Their theories are all mental and my theories all come from and are generated from my physical linear displacement experiences with space-time warps. They may have witnessed subatomic particles using quantum mechanics to discover microscopic evidence of time warps, but none have had physical experiences like me. None of the scientists have ever ventured into the impossible like me. My friend Dr. Arthur C. Clarke's famous quote indicates why I have discovered my theories: "The universe is not only stranger than we can imagine, it is stranger than we can imagine and the only way to discover the limits of the possible is to go beyond them into the impossible." Since I am the only living person that has flown through a time tunnel vortex, I feel compelled to continue researching and telling people about my experience until my end, especially to the younger generations because it will take a long time to solve this mystery.

What is the reality of what I experienced? With multiple concepts and theories about our Universe, exactly what I experienced remains to be seen. Was it a natural phenomenon? The experience over the Great Bahamas Bank may not have been. It may have been produced by an alien mothership. But the electronic fog that I have encountered twice is definitely a natural occurring phenomenon and it is important that it becomes recognized by mainstream science. Someday it should become well known and be just as familiar to us as thunder, lightning, tornados, and hurricanes.

I believe it will be the younger generations that will discover the mystery of electronic fog and how a space-time warp can take place right here on Earth. Over the years I have communicated with hundreds of young people and given presentations to school classes starting as low as the third grade. They have passed on what I experienced to their friends and their numbers have multiplied. My story has been published in many children's books dating back to the 1980's. Just recently an animated cartoon was made about `my flight for children and another video for young adults and they have been watched on YouTube by over tens of millions of people.

Fate led me into the heart of the Bermuda Triangle mystery and that led me into studying the supernatural and the paranormal to try to find the answer to exactly what I had experienced inside the timestorm. After thousands of years of thought and marvel, science is beginning to understand how our planet will evolve and ultimately die. If we are to survive, we must venture to another planet. Discovering the answer to this mystery will help us to get to that new world by using the time warping power of electronic fog. I hope my research will contribute something to this future mission and save lives in the present. If electronic fog is never proven to be real by mainstream science the mystery of Flight 19 and MH370 will never be solved. Even the mystery of the Bermuda Triangle will remain forever unsolved. It will take our future scientists to discover that it is real.

ABOUT THE AUTHOR

Published Books:

- **"The Fog" (2005)** Llewellyn Publications A Never Before Published Theory of the Bermuda Triangle Phenomenon
- **"Beyond the Bermuda Triangle"** (2017) New Page Books True Encounters with Electronic Fog, Missing Aircraft, and Time Warps
- **Bermuda Triangle Legacy (2018)** Online Publication Crossroad Publisher

Interests Activities

- Yachting and sailing
- Certified Master Scuba Diver
- Master Captain (May 2015)
- Back Country Fishing and Off Shore
- Flying all types of aircraft
- Aerial Photography
- Speed Reading
- Book presentations
- Research Speaker
- Commercial Drone Pilot

Education

- Certified Commercial Drone Pilot (July 2017)
- Coast Guard License: Master Captain (May 2015)

- A.S. Physical Science: Palm Beach State College
- A.A. Building Construction Degree; Palm Beach State College
- Studied at FAU; Business Management courses
- Graduate of Speed Reading Lab; 4000 WPM (over)
- Second Generation Real Estate Developer since 1955 in Palm Beach Florida
- Real Estate Developer for Resorts in the Bahamas since 1965
- Commercial Building Contractor and Developer since 1982 in Islamorada, Florida Keys
- Real Estate Broker Owner of Keys Properties since 1982 in Islamorada, Florida Keys
- Commercial Sales and Property Management/Owner/Builder of Keys Plaza since 1983

Professional History and Business Affiliations

- Member of the Florida Keys Board of Real Estate since 1982
- General Building Contractor's License Florida (1972)
- Real Estate Broker's License since 1972
- Realtor and member of the Florida Keys Board of Realtors since 1982
- Realtor and member of the Palm Beach Association of Realtors since 1995
- Owner and President of Gernon Aviation, Inc. located in Wellington, Florida
- Owner of 82205 Overseas Islamorada LLC : Keys Plaza , Retail shopping plaza (1985 thru current 2019)

Certified Flight Instructor

Active Pilot for 52 years

Commercial Pilot Rating

Helicopter rating

Seaplane rating

Instrument Rating

Planes owned and flown:

(2) A-36 Bonanza

Cessna 210 and Cessna 182,

Sea Ray Amphibian

Piper Cherokee

ACCOMPLISHED TV DOCS AND EPISODES:

- Over 50 (Fifty) Television Documentaries
  (See website: www.ElectronicFog.com)
  Theory of the "Electronic Fog" is now considered to be one of the greatest unsolved mysteries of the Bermuda Triangle.
- National Enquirer Newspaper for best true story related to a Turtle named "Charlie"
  Inside Report T.V. did a reenactment for Turtle "Charlie" story and won for the funniest story of the year

Family

Married fifty (50) years to lovely wife Lynn, 1 daughter, Keely Burton and two grandkids; Kali and Reed Burton.

Address

7200 De Medici Cir

Delray Beach, FL 33446

+1-561-281-5492

EMAIL: BERMUDATRIANGLE@BELLSOUTH.NET

WEBSITE: WWW.Electronicfog.com